Testing Very Big Systems

David M. Marks

McGraw-Hill, Inc.

New York   St. Louis   San Francisco   Auckland   Bogotá   Caracas
Lisbon   London   Madrid   Mexico City   Milan   Montreal   New Delhi
Paris   San Juan   São Paulo   Singapore   Sydney   Tokyo   Toronto

FIRST EDITION
FIRST PRINTING

© 1992 by **Bellcore**.
Published by McGraw-Hill, Inc.

Printed in the United States of America. All rights reserved. The publisher takes no
responsibility for the use of any of the materials or methods described in this book, nor
for the products thereof.

Library of Congress Cataloging-in-Publication Data

Marks, David M.
   Testing very big systems / by David M. Marks.
      p.   cm.
   Includes bibliographical references (p. 179) and index.
   ISBN 0-07-040433-X
   1. Computer software—Testing.   I. Title.
QA76.76.T48M37   1991
005.1'4—dc20   91-33890
                                                            CIP

For information about other McGraw-Hill materials, call 1-800-2-MCGRAW in the U.S.
In other countries call your nearest McGraw-Hill office.

Editor: Gerald T. Papke
Book Editor: Lori Flaherty
Production: Katherine G. Brown
Book Design: Jaclyn J. Boone   TPR4

To Software Testers

Contents

Acknowledgments

I wish to thank Bellcore for providing an environment that gave me an opportunity to write this book. I wish to thank all my colleagues at Bellcore who have worked diligently to advance testing technology. I especially want to thank A. E. Ellis, Jr., G. C. Patton, and S. Dalal for their review and comments and T. Colston for her help with the figures. I also wish to thank Boris Beizer, the series editor, for all his help. But most especially, I want to thank J. P. Marks for all her efforts, without which this book would still be a dream.

Introduction

Testing software systems is a formidable challenge. Testers are expected to do an impossible job in too short a time with too few resources. Testing very big software systems magnifies these inconsistencies. Quality systems must be produced and delivered—regardless of the obstacles.

This book explains how to test very big systems and how to manage the testing of very big systems. Big systems are difficult to test not only because of their size but because the system is continually going through changes during the testing period. Further, determining when to stop testing is a matter of economics. Automating the testing process is the means to surviving the pressures associated with testing very big systems. These issues are usually underplayed in books about testing, but in the case of testing very big systems, it is imperative that they be addressed. *Testing Very Big Systems* can show you how to:

- Plan for testing.
- Estimate testing objectives.
- Divide testing into doable tasks.
- Test effectively.
- Document testing.
- Measure testing.
- Automate testing.

The first and second chapters provide a brief history of software testing as well as introducing big systems to the novice. Even if you are not a tester or a tester of big systems, you can still apply the information found here to test systems of any size.

The third chapter describes the problems associated with testing big systems, and the fourth provides an approach to overcoming the problems. Chapter 5 presents a methodology developed over a 15-year period of testing very big systems. Chapter 6 is devoted to how to control the test process. Test procedures are discussed in chapter 7, and the documentation required for testing in chapter 8. Test metrics, including metrics that have been successfully used to determine when to stop testing, are the subject of chapter 9. Chapters 10 and 11 are devoted to managing testing and providing information to higher management, especially some of the pitfalls to avoid. The last chapter discusses the tools needed to automate testing and the specifications for some of these tools is provided in Appendix D.

Testing Very Big Systems is intended for software people as well as students interested in entering the world of testing very big systems. A tester might derive some fresh ideas and a manager some insights. Although testing real-time systems is not specifically addressed, some of the ideas that are expressed can be used for this purpose.

1

The growth of system test

Software testing has changed significantly over the past 15 years. Until the early seventies, most testing was actually program debugging done by programmers. Today, testing is done by independent test teams using formal methodologies with specific quality objectives.

This chapter provides an overview of the changes testing has undergone, introduces contemporary techniques, and discusses some of the problems associated with testing big systems.

Where it all began

After the first program was written, the first test was conducted. An attempt was made to execute the program. It did not work! It was fixed and run again, and again, it didn't work! The process of running and fixing a program and then running it again was known as debugging. The purpose was to correct the syntax and logic errors to obtain executable code. Notice that debugging did not address whether the code did what it was supposed to do—only that it worked.

Towards the end of the 1950's [GELP88], the cost of systems grew, and it became important to show that the system performed as it was supposed to. Testing became a separate discipline that demonstrated that the system performed as expected. Documented test plans were developed by testers who were usually expected to know by experience how the system should perform. They developed a testing model based on their experience or the documents (if they existed) describing how the user would use the system to determine whether the system worked correctly. Later, requirements became an input to test design so that tracing testcases back to requirements ensured coverage.

Testing was still an activity that took place *after* the system was developed, however. The concept of testing phases was introduced to differentiate between unit test and system test. It is difficult to detect system errors if your purpose is to show the system works. Managers learned that the cost of fixing errors approached the cost of developing the initial system. So, in the late seventies, the focus of testing changed to detecting errors.

During the early eighties, analyzing testing costs showed that the cost of fixing a problem caused by an incorrect specification in the testing period was greater than the cost of fixing it in the specification period. Consequently, testing evolved into an activity that takes place throughout a system's life-cycle. One such approach is called, Validation, Verification and Testing (VV&T).

The economics of testing

Testing is not a revenue-producing activity; it is a revenue-preserving activity. A well-tested system will work as customers expect and will have fewer problems in the field, therefore costing less to maintain because fewer fixes will be required. This is the economic reason for testing. Conversely, testing activities can encompass up to half of the effort spent on a system through its life span and must be tightly controlled fiscally. Economics is a driving factor in many testing decisions. Economics is a major factor in when to test, where to test, and how long to test. Testers must consciously balance between testing enough and testing too much.

Testing throughout the development life cycle

The concept of testing throughout the life cycle is referred to as Validation, Verification, and Testing (VV&T). This section discusses VV&T and how to apply it through the development life cycle.

Introducing VV&T

Validation, Verification, and Testing (VV&T) processes are conducted throughout a system's development life cycle [BRAN80]. VV&T is not a single technique but a combination of testing and analysis techniques to validate and verify the system. But before proceeding, it is a good idea to ensure that you completely understand these terms:

Validation A determination of the correctness of the final product produced by a development project with respect to the user's needs and requirements. It is usually done at each stage of the software development life cycle.

Verification A demonstration of the consistency, the completeness, and the correctness of the system at each stage and between each stage of the development life cycle.

Testing An examination of the behavior of a system by executing the system on a sample set of data.

In the following sections, which describe specific milestones in the development life cycle, the objectives of VV&T are discussed and some appropriate techniques mentioned.

Requirements

The first step in the software development life cycle is establishing requirements. VV&T should first be applied during this step. The value of ensuring that the requirements are correct, complete, and consistent cannot be overestimated. Fixing an error that has passed to the next stage is significantly more expensive than fixing it before it is passed on.

The preferred method of developing requirements involves using a requirements language. This is so the tools supplied with the language can be used for verification. When requirements are developed this way, it is easy to develop requirements-based tests and to trace requirements to testcases. However, for big systems with short lead times, the requirements are usually a document about three-inches thick written in something other than a requirements language e.g., English.

Verification of the requirements requires a series of internal reviews and end user reviews. It is extremely important for testers to be involved in these reviews to ensure that requirements are testable. One method to ensure testability is to try and generate testcases that exercise the system as defined by the requirements. If these testcases work, the requirements are testable. These testcases will form the core of the testing in later stages.

While the software requirements are being written, the testing organization begins to develop the test requirements. Test requirements include: the testing strategy, methodology, goals, evaluation criteria, and a schedule with observable milestones. Recruiting the staff for an independent test team should begin at this stage so that testers can participate in high-level design reviews.

High-level design reviews

Design reviews are used to verify that a design is consistent and complete and are best represented using an automated design system or language. If this cannot be done, then at least a design technique such as Top Down Design [MILL70], Structured Design [YOUR79], or the Jackson method [JACK75] should be used.

Following a design review, testcases to exercise the functions specified in the design are developed. These testcases are added to those developed in the requirements phase.

Code and review inspections

After an identifiable part of the system has been coded and compiled or assembled without error, the code should be inspected to ensure it adheres to standards and is consistent with the design and eliminates interface mismatches between modules. A method such as described in [ACKE89] should be used. The inspection is a structured walk-through with an audience of invited experts that should be done before any testing occurs. This rule is hard to enforce, however, because developers, as everyone else, do not relish other people finding errors in their work and tend to delay walk-throughs as long as possible.

Unit testing can begin after the walk-through is completed. If unit testing is conducted by someone other than the developer, testcases could be generated in parallel with the coding.

Unit testing

The developer or a tester in the development group executes the testcases against the unit. Coverage is a critical issue and an analyzer should be used to verify at least branch coverage [MILL86, BEIZ90]. Error messages should be reviewed to ensure that the messages are understandable.

In addition to coverage analyzers, there are other tools that can be used during unit testing. If the unit cannot be tested stand-alone, drivers (software that provides the same environment for the unit that the system does) must be written. If the unit needs information from a yet-to-be-coded part of the system, a stub (software that responds in a prescribed manner to simulate other software or hardware) must be written. All discrepancies found during unit testing should be recorded and tracked to ensure that problems are corrected before integration testing starts.

Integration testing

Logically related units that were previously tested separately are tested together during integration testing. These units support functions that are now tested independently. An integration test plan defines the testing activities. Previously defined testcases are executed. Problems are recorded and tracked. Problem identification and isolation is much harder than in unit testing because interface mismatch problems tend to surface in this phase. It is effective to have the people who wrote the requirements conduct the integration testing. The results of integration testing should be documented and the testcases documented and saved for the next release. The testcases should also be passed to the system testing phase.

System testing

During system testing, the entire system (product) is tested. A system test plan describes the activities, dependencies, responsibilities, and schedule. Previously developed testcases are executed and the results analyzed. Any discrepancies are recorded, fixed, and retested.

System testing is usually conducted by an independent team of testers. By reporting to separate managers, system testers can get some support for fixing errors in-house rather than shipping on time and fixing the errors in the field. The trade-offs involved are not clean-cut because the cost of missing a market window can only be estimated. System testing must verify that the product operates as documented, interfaces correctly with other systems, performs as required, and satisfies the user's needs. Automated testing tools are required because verifying that the system has not lost capabilities supported in prior releases cannot be done economically without automated testing.

A test results report is issued at the conclusion of system testing. This document summarizes any open discrepancies and serves as a feedback loop to improve the testing process.

Acceptance testing

Acceptance testing ensures that the system meets customer needs. An acceptance test plan describes the testing activities. The acceptance test team is composed of customer testers with a liaison to the development and test organizations. When a new feature in the release alters the previous standard flow of operations, acceptance testers will focus on determining the new standard operations flow. Discrepancies identified during acceptance testing are reported, recorded, fixed, and retested. One result of acceptance testing is a recommendation to accept or reject the system. Acceptance testing can be conducted at the developers installation by end users who can make the final decision on whether to ship or not.

Testing the next release

Shipping the release does not mean the end of testing activities until the arrival of the next release. Any automated testcases that were not finished because of lack of time, should be completed. The contents of the regression test package is completed by adding testcases for new functions and removing any redundant testcases. An ongoing activity is analyzing any field-reported problems to determine if any changes are needed in testing. The analysis should identify specific recommendations for changes in typical data, user work flow, and suggested ways to plug holes in the test process.

Testing and the life cycle

Testing a big system cannot be an add-on activity after the development is completed. To achieve a reasonable level of coverage, planning must start at the same time that requirements are being written. Testing continues throughout the software life cycle and does not stop when the release is shipped. Big systems have many versions and the test organization is usually required to test multiple releases simultaneously.

Purpose/quality objectives

The purpose of testing has evolved from debugging to demonstrating that the system works and determining how well the system meets the user's needs. Testers must find discrepancies within the time and resource constraints. After finding discrepancies, testers must determine their severity—their impact on testing and their impact on the user. After discrepancies are corrected, testers verify that the discrepancies are gone. Discrepancies include any instance of the system not performing as documented, any instance of nonconformance to standards, and any instance of the system not meeting the testers perception, based on experience, of the user's needs. Discrepancy identification and correction improves the quality of the system. The testing organization's responsibility is to measure the system's quality through discrepancy identification.

Organization

The testing organization can be part of the development organization or can be independent. Usually, organization for a big system groups developers by a specific function. Even if testers are part of this group, other testers must be grouped together to test the system resulting from combining the functions developed by separate groups. Let's call this group the "system testing group." If the system testers report to a manager who is of equal level with the manager responsible for system development, the testers can be considered independent testers.

Independence is necessary to separate discrepancy identification from discrepancy correction. If both activities are one manager's responsibility, the objectivity needed to determine if a testcase result should be considered as a discrepancy or not is hard to achieve. The manager is constantly at odds because if a discrepancy can be ignored, then the software does not have to be corrected.

Testers can be objective if they are independent. The tester's vested interest in confirming that an unexpected result should always be a discrepancy is a balance to the developer's vested interest that the system works correctly. Often, the investigation of discrepancies results in the attitude of "If you think this is wrong, then prove it!" Showing a seemingly incorrect result is not enough to overcome this attitude, and testers must show in a step-by-step progression exactly how the result was obtained.

The interface between testers and developers is fragile. Testers need developers to confirm that unexpected results are discrepancies. Developers need testers to verify that the system satisfies the user's needs. Friction at this interface can cause testers to become disillusioned, often showing that the system works instead of showing that the system meets the user's needs.

Summary

Testing changed from debugging to VV&T throughout the software life cycle.

The purpose of testing has evolved to showing that the software meets quality objectives of the users.

Testers have the responsibility to prove that the problems they discover are "really" problems.

Economics is a limiting factor in testing.

If too much time is spent testing, the window of opportunity is missed. If too little time is spent testing, quality may be compromised.

2

Introduction to big systems

Big systems are developed to solve big problems and can be characterized by attributes other than size alone. For example, big systems are expensive to develop and maintain, therefore, the users have high expectations about the functionality and quality of the system. A big system is also complex and the system's interfaces with external systems expands the scope of test activities.

What is a big system?

A hypothetical big system will be introduced to serve as a model in later chapters. But first, what is a big system? For purposes of this book, a big system is one that contains more than 3 million lines of source code. [YOUR82] First, let's define small and medium systems.

Small system

A small software system provides a single service, such as an inventory system, an accounting system, or a billing system. Obviously, "small" does not equate to unimportant. Typically, a small system is under 500,000 lines of code.

Medium system

A medium system is one that provides several services. The major difference between a small system and a medium system is the functionality and the integration of the functionality into a user-oriented package. Typically, a medium system is 500,000 to 2 million lines of code.

Big system

A big software system provides many services. The major difference between a medium system and a big system is the complexity of the functions. Complexity can be caused by complex computational algorithms or by complex data relationships. Typically, a big system is more than 3 million lines of code.

Characteristics of a big system

Big systems are developed to address big problems. But the size of the system is only one way to characterize it. Suppose a system that contains 6 million lines of code must be developed. Assuming a programmer can produce 6,000 lines of code a year, it would take 100 staff years to develop this system. The support personnel would also number about 100. On the average, a programmer costs $50,000 per year, other benefits and resources double this figure to $100,000. Therefore, the system costs about $10 million for programming, and doubling this for support personnel, it totals $20 million. Big systems are expensive! Because of the large initial expense, big systems must be amortized over a long life span. Therefore, another big system characteristic is its long life span.

Because of this long life span, big systems undergo significant enhancement rather than replacement. Specific parts of the system might be completely rewritten, but only within the existing architecture. They are also continually being functionally upgraded as the need arises. Some big systems are used by many independent customers. These systems must be flexible enough to be used by organizations that are structurally different and might require that the system have independent paths that support a diverse user community.

A system can also be considered big if the system's databases are numerous and interrelated. Testing a complex system requires that system functions be verified many times. For instance, the function of adding a simple item might cause the addition of records in 10 related databases. To verify the add function, will then require 10 verifications, almost as if there were 10 independent add functions.

For example, a system, call it A, can be considered big if it is an independent but integral part of a bigger system, called B. The reason is that testing system A can require that all the interfacing systems in B also be present, thus creating a big system. NOTE: stubs that provide the responses expected can be used for part of this testing but live testing with the interfacing system must be done.

A system could also be big because of the complexity of the problem it is solving. For example, a system using linear programming techniques to determine the optimum routing for trucks presents a bigger testing problem than one that routes trucks from one location to another without optimizing.

Big systems address large, important problems. Otherwise, their cost would not be economically justifiable. The problems are part of the business and must be solved. Examples of big systems are:

- A system that analyzes nuclear reactors and determines how they act over their life.
- A system that guides the space shuttle.
- The system that is used by telephone companies to inventory, plan, and design the millions of telephone circuits connecting central telephone offices.

Why is it difficult to test a big system? Obviously, size is one factor but other factors also contribute to making testing difficult. Some of the factors that affect testing are addressed in the following sections.

Customer expectations

People who buy big systems have high expectations of functionality and quality. They have spent a large amount of money for an initial version. The customers start using the first version of the system and discover that some things are not as they expected. Some of these differences are caused by misinterpretation of the requirements, others by incomplete requirements.

The user is at first annoyed. After spending so much money, why can't they get it right? System developers must expect this initial reaction and be prepared for it. The user's second thought is, "I want this corrected." In a well-organized project, they submit a Software Modification Request (SMR) to the system developers. Some of these SMRs address the misinterpretation of the requirements. These are called maintenance SMRs. Other SMRs address incomplete requirements and are called enhancement SMRs.

As the system matures, SMRs are written for other reasons. Enhancement SMRs are issued to request increased functionality. Some of these enhancements affect how the users interact with the system. The users change the way they use the system because their organizations change and different people may begin to use the system because they need the data it contains. These differences sometimes uncover new problems that must be corrected and maintenance SMRs are issued to request these modifications.

Customers need all requested maintenance SMRs corrected as soon as possible. The customer's initial expectation is that these SMRs will be resolved in the next release. The development organization must negotiate with the customer to determine priorities if the requested number of changes are more than can be handled in one release. The proper mix of enhancement SMRs and maintenance SMRs is always an issue in determining the release content.

Customers and developers must negotiate the release content for a feature level to satisfy the customer's need for both new features and maintenance and the developer's need for a timely introduction of functionality. Some enhancement SMRs require several releases and changes in many areas before the enhancement is completed. The progression of implementing these SMRs should not be interrupted without understanding the impact.

Customers want new functionality and maintenance but they need a stable system. The latest version of the system must have fewer problems than the previous version and the changes must help the users without requiring much retraining. The user's needs are of utmost importance! When the latest release does not support capabilities that were supported in the previous release, the impact on users is tremendous. Therefore, testers must ensure that all previous capabilities are still supported.

Databases as a company asset

The databases of a big system are also usually big. Reorganizing these databases, as new features are added, can take many hours or days. The time needed for database reorganization is thus another measure of a big systems size and complexity. Systems that handle millions of records are usual. To prevent these databases from becoming too big, they are carefully designed to minimize data redundancy. As the system is used over the years, the data contained in these databases becomes unique and irreplaceable.

The data cannot be recreated by using paper records because the system's daily paper output represents only a subset of the data in the databases. Also, paper output is usually incomplete. Because of the amount and complexity of the data, it can no longer be handled by traditional paper and pencil methods. It takes on a life of its own and can only be manipulated by computer software.

The system data can be accessed by other software, for instance, Mark IV report writing programs, which allows the user to create customized reports or on-line query systems. The combination of the unique database contents and specialized reporting provides the user with an irreplaceable source of information. The user depends on this source and views it as a company asset. The testers must ensure that data is never destroyed by a faulty program.

Interfaces

Big systems often have many interfaces with other systems. Sometimes, a system has grown to be large because previously independent systems that had interfaces have been combined. Other times, a system supplies/receives data from an external system via an on-line link or equivalent. Testing these interfaces presents several significant problems. One way to approach the testing, is to incorporate all interfacing systems into the test environment. Unfortunately, this complicates the software configuration control enormously. Another way is to simulate or emulate any interface that cannot be easily tested. There are two questions that interface testers must address. First is how to test? And second is when to test? Obviously, if the software directly related to an interface changes, the interface must be tested. Less obvious is when the software that generates data that is sent across the interface changes, the interface must be tested.

Customers have special expectations about interfaces. If interfaces are changed, customers require early notification so that the external impacts can be addressed. Other systems could require change and existing procedures require modification. Therefore, testers must ensure that when the system goes into production, the interfaces will work. Some interfaces are paper interfaces, that is data on a report from one system is used as data to be entered into another system. From the customer's perspective, this interface is just as important as an on-line interface. If the report changes, the interface is affected.

Complexity

Testing big systems is more complex than testing smaller systems because more testcases are needed to provide coverage. There are two aspects to complexity: computational complexity and system flow complexity.

For a system that is computationally complex, testcases and data must be included for each unique condition. While this is also true for less complex systems, the size of the data sample is much larger for a complex system. For example, a nuclear reactor analysis system uses iteration to converge to an approximate answer. By changing the composition of elements in the reactor or their location in the reactor model, the answer will change. The tester must choose a known configuration and other variations to verify that the system is functioning correctly. Less complex systems provide a deterministic answer and, therefore, require fewer testcases.

Testing a system with a complex flow is similar to testing a program with many paths. The more complexity, the more the testcases needed to verify that the system functions correctly.

System flow complexity is defined as the condition where there are logical paths through the system that execute subsystems in different sequences. This condition exists when a system is flexible to satisfy the user's organizational structure. For instance, in a system that provided a personnel subsystem and a payroll subsystem, one user might enter an employee in payroll first and personnel second while another user might do it in the reverse order. If the system allows both paths, the tester needs two testcases. Conversely, if it was mandatory to enter personnel first, only one testcase would be necessary. The tester must use the user's operations flow as a guide to testcase development.

Criticality

Correctness of software can be critical to a business' success or failure. Serious legal questions can arise when systems that control such things as a nuclear reactor, life-support equipment, or fund transfers fail. Software can also be critical if the business depends on the operation of the software. The funds transfer system must operate or the bank might not meet their obligations and fail. Imagine the losses if the software system that is controlling an assembly line suddenly fails because of a software problem. The bigger the system, the bigger the failure.

Critical software must be as error free as possible. Testers must verify all normal operations with as many conceivable data values as possible. The testers must also verify error-recovery operations.

The IAM system

To illustrate the test methodology in the following chapters, let us hypothesize a system. The Inventory and Accounting system for Manufacture (IAM) is a big database system with batch and on-line functions. IAM has been successfully marketed and is currently in production in more than 50 sites. The system costs each company $250,000 per year. The cost covers maintenance to correcting errors. The success of the IAM system is based on its ability to manage a manufacturing business while supplying both new functions and upgrades or corrections to existing functions.

The IAM system is composed of independent components so that any component can be revised without major impact to other components. The components and a brief description of each follows:

order tracking   Enters orders and tracks the status (e.g., waiting for inventory, being processed, in shipping completed, paid, overdue).

process control   Determines the inventory necessary for an order and the people needed and assigns them to the order. Determines what equipment is needed and schedules the order onto the equipment. Tracks the order through the plant.

inventory   Tracks inventory on hand, orders when inventory goes below threshold value, interfaces with future market system (outside IAM) to place large bulk orders.

database maintenance   Provides the capability to correct problems in the databases using a low-level, user-friendly language.

personnel   Lists employees and the information associated with each—pay (from payroll), training information, health, etc. Schedules training for employees.

billing and shipping   Bills the customer for the merchandise and determines whether a company truck or an independent trucker will deliver the order.

payroll   Pays employees based on hours and skill level of work during the period (workers can work at several different levels during any pay period).

accounts payable   Pays the bills that the company owes to the suppliers.

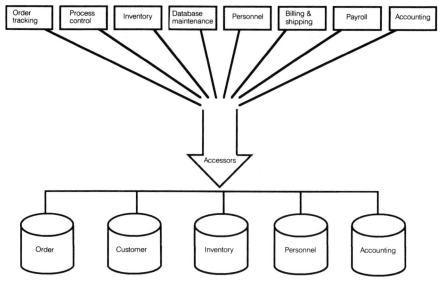

Fig. 2-1. Components of IAM.

These components have been integrated with the least amount of redundant data. The three major information flows in the system are:

1. How an order for an item, from input of the order to shipping the completed item, is handled.
2. The control process that gets the material, people, and equipment needed to fulfill the order.
3. How new inventory is ordered and how suppliers are paid for goods received.

Summary

Many systems today are considered big systems, some are big from the start, others evolve by combining smaller systems. The users of big systems are among the more sophisticated computer users and expect more functionality and high quality for their large investment.

Big system's databases contain a unique collection of information that, after years of use, become an invaluable asset that must be protected from corruption by new or modified software.

Big systems typically interface with other systems, some of which are independently developed. These interfaces, both electronic and via paper, are critical and must be operational.

A sample system, IAM, will help to illustrate testing techniques in later chapters.

3

Problems in testing
big systems

Big systems are difficult to test for many reasons. While it has been proven that complete testing is impossible, enough testing must be done to produce a high-quality system. [MANN78] To satisfy user expectations, systems are produced rapidly and more than one version of the system is in use in the field at any time. As testing progresses and discrepancies are identified, the repair and retest cycle is a cause of concern. Other areas of concern are the lack of documentation, the time required to test with reasonable confidence, and the problem of how to decide when to stop testing.

Impossibility

"Complete" testing, depending on how it is defined, could possibly require infinite time, but would certainly require years. Because it is economically not feasible to spend years testing a software system, it can be deemed as impossible to completely test a system within the reasonable time period normally allotted to testing in the software development life cycle.

Development life cycle

Users of big systems often pay for the system annually and, therefore, expect to receive improved functionality each year. A big system can always be expanded to provide a previously unavailable function if enough money is available. There is an underlying problem, however, in that funding is annual and supports a specific set of improvements called features. Some features might take longer than one year to implement. In this case, either the customer supports a longer cycle or feature development is split into multiple observable phases. The time for each part of the development life cycle for a typical feature of a big system might be allotted as follows:

Requirements—three months   Three months is not enough time for complete requirements development but it is shown on project schedules as only three months. Actually, requirements development starts when a feature is in the proposal stage. Before the customers can decide whether to approve or reject a specific feature, they need preliminary requirements. Thus, the time to create preliminary and final requirements could last six to nine months.

Design and implementation—six months   The six-month figure assumes that the development organization is already in place and that the system architecture supports the new feature without major changes. The implementation phase includes unit testing and multi-unit testing (the combination of logically related units that support one feature).

Independent testing—three months   During this interval, the feature is tested within the framework of the system. Also during this phase, called "product test" by some, the entire collection of software that will be released to the customer is tested for both new functionality, old functionality and, if appropriate, the ability to interface correctly with other systems.

Included in FIG. 3-1 are the activities packaging, getting the system together and ready for delivery, and installation, making the system operational at the user's location. By overlapping these activities, a 12-month cycle can be achieved. This year-long cycle ensures that software systems are upgraded at least once a year.

Release strategy

Shipping the newest release of a system, with all its features on schedule, is the reason for a phased, planned development cycle. However, the customer has many reasons for wanting to get more frequent releases or less frequent releases. Releases can be divided into three categories: major releases, minor releases, and maintenance releases.

Major releases contain new features that could require structural changes (i.e., a new database, modifications to existing databases, a new interface). The development cycle is one year and the test period is three months.

Minor releases contain new features as modifications to existing functionality and require no structural changes. The development cycle is four months and the test period is one month.

Maintenance releases fix problems in the current software and are developed as problems are discovered. The test period is one week.

Major releases have a large impact on a customer's business. Personnel need training to use the new software and retraining to use changed software. However, a major release usually includes capabilities that increase productiv-

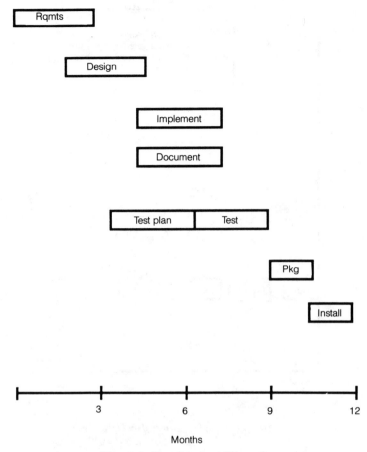

Fig. 3-1. Development life cycle.

ity through increased performance or new functionality. Customers with big systems want the improvements but might not be able to train and/or retrain their personnel immediately or be able to install the software until hardware upgrades are made. This leads to multiple major releases being active in the field. This presents a significant testing requirement, the capability to test two or three major releases concurrently.

For example, let's say one customer is using production release version 9, another is using version 10, and version 11 is being tested before release. If both customers encounter problems that must be corrected in maintenance releases or if the next minor or maintenance release of the system is scheduled, the test groups could be faced with two or three releases that require testing at the same time. Figure 3-2 shows a possible combined schedule for releases 9, 10, and 11. Note the overlap in months 10 and 23.

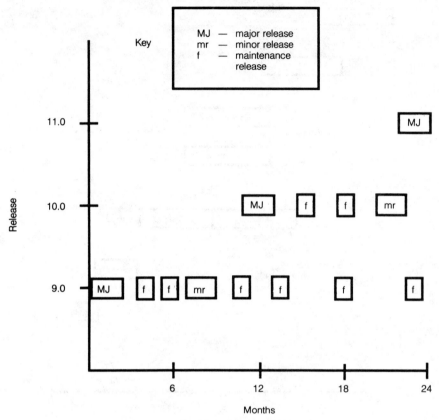

Fig. 3-2. Overlapping test intervals.

Release building

To support multiple releases required by big system users, it is necessary to have a Software Configuration Control System (SCCS). A SCCS must allow multiple versions of source code to be obtained and modified concurrently. The SCCS must provide a history of changes made to each version of the source. The SCCS also must have the ability to reconstruct prior releases, at any level, to provide a backup for disaster recovery. For a big system, 100,000 items might be controlled by the SCCS and 10,000 of these could be modified for a major release. Items that should be under change control include programs, macros, and database definitions. Coordinating and bringing all these items together into a release is a process called "building a release." It includes: compiling (from source to object), assembling, constructing files that contain reference data, documentation, etc. Building a release requires significant resources: people, software tools, hardware, and computer time. Also, as the environment changes, for instance, if a new language is used, the software tools must be updated.

Testing difficulties

Some of the reasons that big systems are difficult to test are the same as the reasons that big systems are difficult to develop. Specifically, several difficulties that are encountered when developing big systems are the need to:

1. Plan in great detail.
2. Start development despite the late availability of requirements.
3. Finalize requirements and design so that architectural changes are eliminated.
4. Close communication between development groups even though they report to different managers.
5. Develop a configuration control system for software, requirements, documentation, etc.

Testing is also difficult for reasons that are unique to testing—the lack of documentation, small number of testers, and the time it takes to test. Also, the question of "When is testing ever complete," as well as the lack of value placed on testing, compounds the testing process.

Lack of documentation

One subgoal of testing is to verify that the system performs as specified in the documentation furnished to the user (major goal is to verify that the system performs as required). User documents are developed, at best, in parallel with the software and usually are completed after the software is available for system test. Therefore, the documentation is not available for use in planning and not available during the first part of testing. This makes it difficult to verify that the system performs as documented.

When the software change is small, as in a minor or maintenance release, the testing cycle is shorter than the testing cycle for a major release. To test efficiently within this shorter period, testing concentrates on the areas affected by the changes. This seems straightforward but often changes made in one area impact other areas. For example, suppose a problem was corrected in the order-tracking component of the fictitious IAM system mentioned in chapter 3. Orders that had a future billing date that crossed the year boundary were incorrectly purged at the end of the year. The fix ensured that these orders would remain in the database. Testing this fix also requires that the billing and shipping components be tested to ensure that they operate correctly on this type of order.

Documentation is needed to address these dependencies but such documentation is difficult to produce and maintain. The lack of this documentation means there is a greater risk that testing for the impact of a change might not be as thorough as required.

Number of testers is small

Generally, testing uses half of the development time for a system. Understandably, one might expect that half of the project personnel would be testing. Typically, however, only one-tenth of the personnel are assigned test responsibilities. This small number of testers means that the test organization cannot respond as quickly to product changes as would be possible with more people.

Constant retesting

A series of testcases are run at the start of a test period, some of which discover faults. These are reported and fixed. Because some fixes are more important and some can be corrected quicker, the fixes are ready to be tested at different times. One way to approach this time disparity is to batch the fixes and deliver the software only at prespecified intervals using a process called incremental delivery. The major drawback to this approach is that often when a fault is fixed, the next fault is immediately discovered. This process is like peeling an onion and the "real" testing starts only after several layers have been peeled away.

When testing a big system that contains many changes, the number of increments needed depends on the real-time results of testing and cannot be predicted. Incremental delivery planning is traditionally based on initial software development scheduling rather than on the reality of the incremental release of fixes. The test interval is usually too short to wait for fixes. Therefore, fixes are accepted into test as soon as they are available (which assumes prior unit and multiunit level testing). Immediate delivery of fixes shortens the period to retest but it also means that some previous testing might not be valid and must be rerun. The only way to cope with a volatile environment such as this is to develop mechanized testcases that can easily be rerun to reverify that a fix did not actually impact another area of the system.

There is no such thing as a quick test!

Often during a project, development managers ask the test organization to do a quick test of a "small fix." There is "no" quick test for a big system! What can be done is to quickly verify that the fix corrects the problem that was found. However, that test does not verify that everything else was unaffected. Regression tests, which do ensure that the system operates as it did before the fix, do not happen quickly. A big system requires many tests and requires significant test time.

When are you through?

There are an infinite number of possible testcases that could be used to test a big system. A selected subset is used and their execution discovers a group of problems. These problems are fixed, the software is retested, and more problems are found and the fix-retest-find new problems cycle continues. The difficulty is: when should testing stop? One approach is to stop testing after the last problem is discovered. This ignores a scheduled completion date, however. Another approach is to stop testing when the last planned testcase has been successfully completed. This ignores any consideration about what problems have not been found, however. One factor that can help is to have a test coverage metric that measures how well the system was tested. Yet another approach is to stop testing when the cost of finding the next problem exceeds the cost of the problem's potential impact on the customer and the cost of fixing the problem when the system is in the field. A difficulty with this approach is that the calculation includes two items that are difficult to quantify: the time testers need to find the problems and the risk associated with the potential impact on the customer. The decision to stop testing is always a judgment.

Testers are undervalued

The decision to stop testing explicitly assumes that the testing intensity is constant throughout the test interval. For this to be true, the testers must be highly motivated, which means that the testers must believe that testing is rewarding both to themselves financially and to the product in terms of added value. With the current drive for higher-quality products, it is easy to see the reward in terms of a higher-quality product but the financial reward is lagging. Even if the reward system is based on value added, when the contribution of the tester, who measures the quality of the product, is weighted against the contribution of the developer, who produces the product, the tester's contribution appears less. One way to improve the testers apparent contribution is to encourage testers to contribute in other ways along with testing. Some examples are: user training, documentation, installation at the customers site, and interaction with the customers on problem resolution.

Summary

In addition to the sheer size of a system, other factors make testing big systems difficult. While each release of the system follows a planned progression of steps called the "software development life cycle," the steady-state reality is that there are several releases in the field concurrently. Configuration control is absolutely mandatory in such a volatile environment.

Other testing difficulties occur because of:

- A lack of testing-oriented documentation that correlates changes in one area to impacts in other areas.

- The small number of testers.
- Testers must retest previously tested functionality as testers find discrepancies and the software is modified during the test cycle to fix the discrepancies or to increase functionality,

Testing is time-consuming. Because of the above difficulties, big system testing is heavily dependent on regression testing to prove that the old functionality works correctly. Regression testing takes time.

A decision to stop testing risks not finding the next critical discrepancy. There is less risk if the testing process continues to the point where the cost of finding the next discrepancy is more than the cost to fix the discrepancy in the field.

Talented testers need opportunities to demonstrate their skills in roles other than the traditional test role.

4

Testing approach

In spite of the difficulties of testing, it is an important criteria to designing, upgrading, and maintaining any system because it measures the quality of the system and provides a final opportunity for discrepancies to be found and corrected before the software product is delivered to the customer. Activities that should make testing a system easier, such as planning, documentation, economic analysis, and content control, often provide ambiguous information that must be clarified in a "testing approach." Other questions that must be considered in the "testing approach" include:

- When to automate testcases.
- When to test manually.
- What to retest at the end of the test period.

The test process

The best approach to testing big systems is to put a test process in place and follow it. The test process must include:

- Planning for testing.
- Defining testing objectives.
- Designing testcases.
- Preparing for testing.
- Executing testcases.
- Evaluating software.
- Evaluating testcases.

The first step in developing a testing approach is to partition the testing effort into separate testing phases. Each phase then has the primary purpose of demonstrating that the software conforms to requirements and of finding discrepancies before they reach the next phase of testing. Phases can be determined by: who does the testing, where the testing is done, how much software is tested, and what is the emphasis of the testing. The following phases have been developed after many years of experience with many products.

Unit test is done by a single developer on an entity no smaller than the source identified as a single module in the Software Configuration Control System (SCCS). Unit testing is usually performed in an isolated environment, possibly a workstation. Unit test will test as much as possible and will focus on input domain coverage, output range coverage, branch or possibly statement coverage, and error handling.

Multi-unit test is done by a development team on collections of logically related units. Multi-unit testing is usually performed in a development environment but could be done in the test or user environment. Multi-unit testing should focus on externally observable old and new features, input domain and output range coverage, and unit-to-unit interface coverage.

Product test is done by a team independent of the developers on the complete set of software that is to be delivered to the user. Product test begins after multi-unit testing is completed and is done in the test environment. Product testers must verify that the product will meet the users expectations of product quality—functionality, reliability, availability, maintainability, security, and usability. The product tester will focus on user-observable functions, input domain and output range coverage, and the sanity of interfaces to other products.

Inter-product test is done by an independent team of testers on a collection of products in the user environment, which, if not available, is simulated. Inter-product testing can be performed when product testing determines that the product is stable. The focus of inter-product testing is user-observable functions and coverage of product-to-product interfaces.

Evaluation testing is done by a team, independent of developers, on the deliverable product in the test environment after product testing is completed. Evaluation testing is user oriented and has the purpose of recommending if the product is ready to ship. The focus of this testing is normal operations (not boundary conditions) of established functions.

Testing process activities that apply to each of these phases are discussed in the following section.

Planning for testing

Planning for testing must begin at the start of the project, as soon as the requirements are understood. The project plan must include a section on testing. This is called the project test plan. Each phase of testing also has its own test plan. These test plans address:

Responsibilities   Who will do the testing?
Activities   What specific activities will be performed?
Productivity   What will the output of these activities produce?
Standards   What are the policies and guidelines to follow?
Technique   How to do the testing?
Metrics   What metrics will be used to measure the quality?
Tools   What will be used or developed to do the testing?
Cost   How much will it cost?
Risk   What can go wrong?

Defining test objectives

In the test plan, various test phases and types of testing can be specified. For each of these, a specific objective must be defined for each type of coverage. The objective must be measurable so achievement can be determined. It cannot be a subjective statement such as "Test completely." An example of an achievable objective is, "Verify that at least 95 percent of the code is executed!" Objectives should be defined for each phase for the following types of testing.

Basic function   Tests the basic sanity of the software and is useful criteria
                   for transition to the next phase of testing.
New function   Tests the quality of new functions with a user orientation.
Old function   Tests the ability of the software to continue to perform
                   functions that it performed in previous releases. It is often
                   called regression testing.

The type of coverage specifies:

Input domain coverage   How large a sample of the input should be
                          selected.
Output range coverage   How much of the output range should be pro-
                          vided.
Structure coverage   The quantity of the structural element exercised
                          by the testcases, e.g., 95 percent of the branches.
Error response   The coverage of error-handling software.
Interfaces   The interface coverage to other subsystems and
                          systems.

Designing testcases

The test plan and the test objectives are the requirements for designing test-cases. This activity is performed for each phase of testing. Designing test-cases encourages: organizing test objectives into sets—partitions the testing effort into manageable chunks; selecting sample input—picks specific data values to be used; determining output range—predicts the specific outputs that will be generated; tracing testcases to requirements—develops a list of testcases for each of the requirements that will be tested. This should be a two-way correlation so that if a requirement changes, the affected testcases can be changed or executed.

Preparing for testing

Using the information provided by the testcase design, testers must obtain:

- Input—acquire or generate the input needed by the testcases.
- Environment—develop the environment to be used for testing, including computer hardware and communication lines.
- Tools—develop or get the necessary test drivers and output capture mechanisms.
- Scaffolding—develop or get the stubs or drivers needed.
- Testcases—code testcase scenarios and run testcase generators to develop the testcases.

Executing testcases

When the software is available, the testcases can be executed. Execution will produce the following:

- Output that can be captured for comparison and/or other analysis.
- Reports that cover requirements, structure, and testcase execution metrics.
- Additional testcases that can be added and executed if reports indicate coverage deficiencies.
- Discrepancy reports to analyze testcases that fail. Testcases that indicate that the software is not performing correctly will result in discrepancy reports. Testcases that were designed/developed incorrectly will be corrected and rerun.

Evaluating software

Using the results of the executed testcases and the predefined pass/fail criteria for the software (defined in the test plan), you can determine if the software meets expectations. Before the determination is completed, however, the discrepancy reports must be analyzed. If any problems are still open, they must be reviewed. If the impact upon the user is small, the system can be released. If the impact is significant, as determined by user agreement, then the system must be corrected.

Evaluating testcases

Evaluating testcases is done throughout the testing period and continues when the software is in the field. The evaluation addresses testcase coverage, redundancy elimination, and data sample revision.

Do testcases provide the coverage needed? For instance, if you want to achieve a given level of code coverage and the results of executing the set of testcases does not exercise sufficient code, then additional testcases must be added.

The set of testcases used for regression keeps growing as testcases are added for new functionality. One type of evaluation is selecting which testcases to add to the regression package. Another type is the periodic review of the entire regression package to eliminate redundant testcases.

After the software is in the field and problems are reported, the data sample must be revised if any problem was not discovered in test because the data sample did not include it.

Ambiguities in the test process

There is no doubt that a test process is needed. However, when a single step is analyzed separately, ambiguities become obvious. Slavish devotion to a maxim such as, "Verify that the software conforms to every document," could be setting unattainable goals. Several areas of ambiguity are discussed in the following section.

Planning

The importance of planning for testing cannot be overemphasized. But because of the volatile nature of big systems, no amount of planning can address every possibility. Each release of the system is different and most planning extrapolates from one release to the next. Using history is better than reinventing the wheel but there can always be the instance that doesn't fit. Even after years of experience with the same system, one release can behave differently. Two examples of where planning can break down are start-up costs and automated testcases.

Start-up costs Before testing begins, testing tools must be updated. This might mean only defining new libraries and copying testcases but it could mean developing and debugging a new tool. Testcases might need updating because the system's functionality changed. When the system is integrated and tested for the first time, a flurry of problems classed as start-up problems, are identified. These could include: expected software that was developed but mistakenly not delivered; software mistakenly delivered at an incorrect level; or a change made that is not tested sufficiently so that the software doesn't work.

These problems must be identified and corrected before effective testing can begin. Planning can recognize that the start-up phase will occur, and by using historical data, hypothesize how long it will last. Just keep in mind that for any specific release, this phase could last longer when too many unexpected problems occur.

Automated testcases   The goal is to automate all testcases before the end of the test period. This requires that the testers learn the system, identify the data, and develop the automated testcase. Often, automation is not achieved because there is not enough time provided in the test plan. Testers learn the system by manually executing testcases. If sufficient time is not provided to automate testcases, manual testing replaces the preferred automated mode.

Planning cannot predict how long it will take a tester to learn a release or to develop automated testcases for each release. Testers usually have mixed responsibilities, which includes testing old software and new software, and their time could be consumed verifying old software. In this case, there would be no time to automate testcases for the new software.

Documentation

A big system has a vast amount of documentation associated with it. The tester could have access to requirements, software designs, software specifications, test plans for earlier phases, test results from earlier phases, and user-oriented documents describing the system and how to use it, but these documents may not be available when the tester needs them. The requirements, for instance, could have changed significantly but had never been updated between the time they were issued and when the tester needs them six months later. User-oriented documentation might only be available during the last month of testing.

The availability and usability of documents must be factored into the test plan. Consequently, the tester must choose what to read and what to ignore from the available documents. The highest-priority documents are those that describe to the user how to use the system. The lowest priority documents are those that describe how the system works internally.

The testers must also generate documents. How much test documentation is necessary? First, a test plan, either formal or informal, is very valuable. Test specifications are useful for transferring test responsibility later. Other documents discussed later in the book can also be profitable. The project test plan should specify the priority for each test document.

Economic analysis

Cost is an important factor when making testing decisions, but determining costs is not easy. One measure of the total cost of testing is to sum up the time people spend during the test intervals. The cost of other resources, such as hardware, must also be factored into the total cost. However, this is not a straightforward summation if multiple releases are being tested concurrently. It also becomes even more complicated if the time people spend at non-test activities such as training, conferences, etc. is eliminated. Another measure of the cost of testing is the cost to identify a discrepancy. This cost is initially high because the rate of finding discrepancies is reduced by: testers who are learning how the new features work; a single problem that blocks further progress; and parts of the system that might not yet be available for testing.

The cost flattens to a constant during the middle of testing and goes up at the end. The rise at the end can be traced to other project demands, required retesting, and prolonged testing.

At the end of the test period, the project's emphasis is shifting to the next release and testing is not as efficient. As the software is modified, it is necessary to retest and to verify that previously working testcases still work. If testing goes on too long, the time to find the next discrepancy increases.

Figure 4-1 shows the change in the cost of identifying a discrepancy during the testing interval.

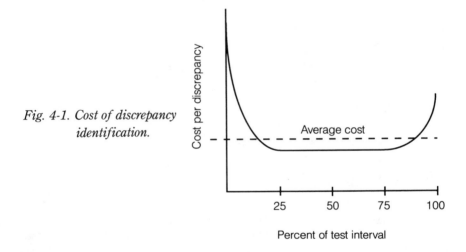

Fig. 4-1. Cost of discrepancy identification.

Content control

Unless the tester is aware of a specific change, it will not be tested. Therefore, controlling a release's content is critical. A software change request (SCR) is needed to identify and describe the change. But changes also arise because of unexpected fixes, cross-area impact, and propagations.

Unexpected fixes arise when a developer changes the code in response to a SCR and finds a previously unidentified bug and corrects it gratuitously. This can change the system's operation without any notification.

A cross-area impact is when a change in one area impacts other areas. For instance, if the format of the order record in a system was changed, it would be necessary to check all the order-tracking software. But it is also necessary to check process control and billing and shipping software that modify the order record.

Propagations occur when a change is made to a prior system release; the change must also be made to the current release under test. If the specific software, in which the change was made in the prior release, is not being changed in the current release, then the changes can propagate (be copied into the current release without impact).

In all of these cases, the content of the release might change without the formal notification of an SCR. These indirect changes must be monitored very closely to ensure that all changes are tested.

Automation

Testcase automation is necessary for big systems and especially essential if the environment is volatile with changes occurring throughout the testing period. Attempting to test the system with only manually executed testcases would require additional test personnel each time a new feature was added. The testers would be totally consumed verifying that all prior testcases still worked! Automation is the only answer. For automated testcases to be effective they must be repeatable, deterministic, and easy to maintain.

If testcases are repeatable, they can be run again without any changes or database initializations. Deterministic means knowing the correct answer before execution so that the testcase (or driver) can check that the right action was taken. Finally, if testacases are easy to maintain, they can be converted from one release to the next with minimum effort.

Well-designed testcases are discussed in further detail in chapter 7.

Manual testing

While automated testing forms the backbone of the testing efforts, some testing is done manually. Manual testing is necessary when the testers do not have time to automate testcases or when testers do not have the knowledge to automate testcases. Manual testing enables testers to become familiar with the system. Testers can also use the results of their manual testing to determine which automated testcases they should develop. No matter how testcases are selected, one important factor that must always be considered is the tester's experience at pinpointing the areas that usually have problems. Until a tester has "played" with a system manually, the system is only a paper system.

The use of manual testing as a way to become familiar with a system should not impede efforts to automate testcases. A minimum effort would be to automate some very basic testcases as soon as possible and leave more complicated testcases for later.

Retesting—size versus scope

Near the end of the test period, the focus of testing must shift from initial testing to retesting. This is the time to verify that the system is ready to ship. Testers must ensure that the system still performs old functions correctly, performs new changes as expected, and performs new functions as required after fixes have been delivered for all known problems. Because it is a big system, there are usually many problems and, thus, many fixes. If all fixes were retested as thoroughly as they were originally tested, there would not be sufficient time. Therefore, regression test packages are used heavily to ensure that changes do not affect old capabilities. Newly automated testcases ensure that new functions perform correctly. Manual testing of broad areas, rather than testing specific fixes, is used to verify changes not covered by automated testcases.

Summary

Testing must be performed in spite of possible difficulties. Planning cannot cover all the contingencies but important decisions such as what to test manually and what to automate must be made early. The flow of documentation into a testing organization is tremendous, and to prevent testers from becoming bogged down by all the paper, a decision must be reached about which documents to read, and thus base testing upon, and which documents to ignore. Also, a decision about what documents to produce and what level of detail is necessary.

To use economic analysis effectively, one must prime the effort with some parameters, such as the cost of testing for one day, a cost that can be hard to define. Testing big systems requires tools scaled to the size of the job. The content of the software must be constantly monitored and controlled. Otherwise, chaos can result. Automated testing, especially for regression purposes, is a necessity. However, manual testing has its place. Because of the volatile nature of software changes coming into the system during the testing period, retesting at the end of the testing period is required. The scope of retest should be defined to cover a broad spectrum in a short time. The way to accomplish this type of testing is to use a testing methodology.

5

Testing methodology

A testing methodology maps testing to practicality. A good testing methodology must be easy to understand and simple to follow so that it can be used by inexperienced testers. Several truisms provide testers of big systems with several guidelines. These guidelines furnish testers with a way to:

- Separate testcases into categories by specific testing tasks.
- Maximize automated testcases.
- Utilize available documentation.

The quality of testing rests upon the coverage provided by testcases. Coverage can be determined in several ways and many different methods are available. The life span of a big system is longer than smaller systems, so a greater value is placed on the automated testcases in the regression test package.

Big system testing truisms

Five truisms that apply specifically to testing big systems are discussed in this section. These truisms provide the basis for the methodology proposed for testing big systems.

Any test is better than no test   If the software is not tested, there is little confidence that it will work correctly. Any software test increases confidence.

The ultimate test never works   Requirements provide the specifications for how a complex software system should work. A complex testcase demonstrating the full complexity of the software will seldom work initially because of software discrepancies. Because this testcase is complex, it is difficult to isolate the specific software faults. Simple testcases are needed to ensure that the system works before attempting complex testcases.

Test where the problems are   Parato analysis has shown that 80 percent of the problems are in 20 percent of the software [POST88A]. Thus, if you find a problem, there is a significant likelihood that another one exists in the same code. If unit testing was not done and the software is in product test, then do unit testing. It's the fastest way to find problems and it will speed up testing.

Use the black-box testing method   Big systems have many piece parts. If you attempt white-box techniques such as code coverage during product testing, the time required for testing increases. During unit testing, code coverage is an appropriate technique because the size of the software being tested is small. As the size of the software being tested increases, the time to execute the necessary testcases to verify coverage increases. The focus of white-box testing is the code rather than the user's view of the system. Gray-box testing addresses the software on a modular basis and is appropriate for multi-unit testing. When applied during product test, however, it also requires too much time. The black-box method is best suited to verify that a system will perform as the user expects.

Automation pays for itself   The life span of a big system is long enough so that the cost becomes relatively small when an automated testcase is spread across many releases. The cost of automation must also be weighed against the cost of testing without automation and the risk of having insufficient test coverage.

Levels

Big projects are handled by dividing the test effort into smaller, more focused sections. The smaller sections are called levels of testing. The levels in one scheme, which have been developed over many years, are time-dependent and have been created to identify specific classes of discrepancies as early in a test cycle as possible. The first level is the basic integrity test.

The basic integrity testing (BIT)

The basic integrity test verifies that the system is grossly correct. It is inefficient to begin testing if the basic elements such as the programs, databases, and screens are not correct. Simpleminded testing, such as invoking each transaction with no data, can uncover many of these discrepancies. BIT testing is always done, sometimes informally, sometimes more formally. Sometimes, BIT testing consists of reading reports to ensure that the elements are correct. Each time a new release of a big system is ready for the next phase of testing, the first batch of tests must verify that further testing is profitable.

Screen testing

The purpose of screen testing is to verify that an on-line system handles input and output correctly. Screen testing is done on a screen-by-screen or format-by-format basis. Many systems use commands rather than screens but testing commands is equivalent to screen testing. Screen testing should be a unit-testing activity whenever possible and repeated in later testing phases. Input or output processing errors should be found early in the test cycle. These errors will cause many testcases to fail because most testcases require screen input and check screen output.

Function testing

The purpose of function testing is to verify that subsystems, or collections of related programs that perform a single function, operate correctly. Function testing should be performed first during multi-unit testing. Function testing addresses module-to-module communication. It is important that these errors be identified and corrected before the system can be effectively tested as a whole.

Workflow testing

The purpose of workflow testing is to verify that the system can perform the necessary user operations. Workflow testing is modeled on a user-usage profile, if it exists, and provides coverage for expected variations in the data or variations in the control path of workflows. Workflow testing depends on the previous levels having been completed because identifying discrepancies becomes difficult when more of the system is exercised in a single testcase.

Characterizing software for testing

Another way to divide the test effort into smaller sections is to separate the software into conveniently testable entities. An example of this separation would be each program function key's operation for an on-line screen (e.g., add, update, etc.) or each batch run's operation (e.g., initialize, update, etc.) for a batch program. The separation process can become creative if no direction is provided, but experienced testers have acquired an ability to intuitively determine a minimum set of testcases to verify software. The knowledge of how to develop these testcases is based on the ability to characterize the software.

What is characterization?

Characterization is the process of determining the similar qualities of entities. When applied to software testing, characterization determines a set of functions that span the range of software capabilities. One example is software that adds a record. A test designed to verify the correct operation of "add" software, can be easily adapted to test other software that "adds." It is easy to extend this principal by forming the union of software characterized by functions to determine the testing requirements for any new software. For instance, if software performs both "add" and "update" functions, then use both the "add" and "update" test specifications to form a new test specification—the union of both.

Why characterize software?

It is easy to define tests for any software if the characterization process produces a small set of software functions to test. Functional characterization should also provide the following benefits:

The ability to use previous tests, with changes, to verify new software   New software will not require new test specifications, only analysis to determine its functional type. Existing testcases can then be changed to provide the required test capabilities. This process is much quicker than the normal design, implement, and debug cycle needed to develop new testcases.

Central control of test specifications   By focusing the most knowledgeable people on a small set of test specifications, well thought-out, thorough, and complete specifications will result. Any change/upgrade to these specifications is applied to the master specification and is used in all new testcases for that software function.

A management tool   If the test specification is developed using test areas and testcases, test managers can assign priorities to each test area and testcase. These priorities can then be used to guide test development. For example, let's take the "add" software function previously discussed. If we assume that three possible testcases would be:

a. add using valid data
b. add using invalid data
c. add using default values,

then the test manager might rate *a* and *b* as high and *c* as low.

A training tool   New testers can begin testing immediately based on a previously defined and prioritized specification. These testers will not have to determine what to test in a vacuum.

How to characterize software

The functional characterization process is done by:

1. Proposing a set of functions from a "testing perspective."
2. Checking that the proposed set of functions addresses each capability of the software collection to be tested.
3. Reducing the number of functions, if possible, by combining members of the proposed set of functions that possess common characteristics.

The term *testing perspective*, used in step 1, bears some explanation because it is a key term in the characterization process. For example, one might assume that one software function would be "calculation." However, using a testing perspective versus a programmer's perspective, one views the software in terms of how to test. From this perspective, the calculation is visible (testable) only after a display or an update to a data set and, thus, "calculation" is not a separate function to be tested.

Testing software functions

The set of software functions to be tested has been used for about seven years to test several hundred batch programs and more than 1,000 on-line transactions without any new functions being required. The seven functions are described both from a user viewpoint and from a testing perspective.

1. Add
2. Communication
3. Delete
4. Interface
5. Query
6. State dependent
7. Update

Add   Add software adds a new record or a new segment to an existing database. The source of the added data could be another system, data subsets already in the system, control cards for a batch run (one obvious case is an initialization of a database), or input from a screen. From a testing prospective, add software must be tested to verify the creation of a destination data set, the addition of an initial record, and the actions caused by duplicate records.

Communication   Communication software deals with communication networks and protocols without knowledge of any external application software. This software typically delivers messages to various destinations, usually hardware devices or data sets used when storing and then forwarding messages. From a testing perspective, communication software must be tested to verify that all hardware destinations, line configurations, and error-recovery processes work as expected.

Delete   Delete software deletes or purges entire records or subordinate records from an existing data set. A typical example is the purge of every record that has already been marked as processed. From a testing perspective, delete software must be tested to verify that no data is destroyed inadvertently, either by direct delete or by deletion of a parent record, (which should be invalid because subordinate records exist) and that deleting records leaves the data set available but empty.

Interface   Interface software communicates to another component subsystem or to an external system. Interface software has knowledge of the target system because the actions performed by interface software depend on software specifically developed for the target system. Interface software can send data, change local data sets, or change the external system's data sets.

The actual communications could be direct or indirect. From a testing perspective, interface software requires that testing be scheduled between both systems and that the testers have knowledge of both systems in order to check results. Testing must verify that each system's data is protected, error recovery works, and system flow tests operate correctly.

Query   Query software accesses the local data sets and presents the results either on a screen or via a report. The format of the resulting display or the data content of the report can be determined by a predefined algorithm or user-specified input. From a testing viewpoint, query software must be tested to verify that data sets are accessed correctly and completely, displays are formatted correctly and completely, security issues are handled, and computational algorithms are correct.

State dependent   State-dependent software acts like a finite state machine in that the current action is dependent on previous actions, or states. An example is a "query" action that could return either:

- A specified record—if no previous queries were issued.
- The "next" record—if a previous query was issued.
- An error message—if no next record was available.

From a testing perspective, state-dependent software must be tested with sequences of actions rather than single actions. Each possible previous state must be tested with valid and invalid successive actions.

Update   Update software changes the data content but does not change the format of an existing record or records in one or more data sets. The data change could be defined by an algorithm. From a testing perspective, update software must be tested to verify that updates do not inadvertently destroy data and that the calculation algorithm is correct.

After characterization, the next step is to determine a set of test items that address all the concerns of testing each software function.

Test areas

Testing is the summation of the execution of all the individual testcases required to verify that a system operates correctly. In this section, items called "testable items" are clarified to avoid any possible bias caused by preconceived notions associated with the word *testcase*. Before developing a long list of testable items, however, let's pause and reflect about the testing function in general.

Testing activities can be divided into four areas: Data set conditioning, input, execution, and result verification. The concept of testing areas is simply a convenient way of classifying testable items. Each test area is described below.

Test area 1—data set conditioning

Data set conditioning verifies that the software being tested can handle various initial conditions in the data set. A few of the initial conditions that should be tested are valid records, invalid records, and record volume.

Valid records are used as a starting point in transactions that change the data set. Invalid records show that the software can overcome bad data. Record volume has enough similar data to force the software down alternate paths, such as more than one page/screen of output.

Test area 2—input

Input area tests determine if input parameters, or combinations of input parameters, are properly handled by the software. Some examples of the input data required are: valid data, invalid data, and default values.

Test area 3—execution

The purpose of tests in the execution area is to verify the software's ability to handle various execution time conditions. Two examples of execution time conditions are hardware errors and software problems.

Hardware errors means software that handles hardware must operate even if the hardware is not in operation. Software problems means software that communicates to, or with, other software must operate even if the other software does not.

Test area 4—result verification

Result verification verifies the results obtained by testing the software. Standards must be adhered to and algorithms must be correct. Examples of result verification are:

- Understandable output messages—messages to the user should be user-oriented.
- Multi-page/screen output—verify that multi-page/screen output is formatted correctly.
- Algorithm correctness—expected input and data set conditioning results should be checked.

Testable items

Test areas are further delineated into a series of testable items. A testable item belongs to only one test area. These items are not to be confused with testcases. A testcase could be composed of one testable item, one testable item from each area, or several testable items from each area. The list of testable items in this section is a good starting point for developing a test plan as it contains most, if not all, of the conditions that must be tested. Individual testers sometimes have additional testable items that must be added to the list, however. The recommended guideline to follow in adding additional items is that most testable items should apply to more than one functional type of software.

Some of these items are applicable to both on-line and batch software. Others are applicable to batch software only. This is denoted by the word "batch" in parenthesis after the test item. Similarly, items applicable to on-line software only will have "on-line" in parenthesis.

Care has been exercised in defining the testable items to phrase each item in a general manner. Note the use of "record" rather than "segment." It is, however, impossible to describe testing without including the testing tools, the on-line/batch interactions, and the database environment. Specific tools that could be used include: DLT0, the IBM-supplied database access facility and RECSELECT, a data set print utility. Testable items are grouped by test area. A complete description of each testable item can be found in appendix A.

Data set conditioning

1. Valid records
2. Invalid records
   - data that will fail validation checks
   - illogical database records
   - illogical application records
   - records that are not integrated
   - record with an invalid destination
3. Every field supplied
4. Minimum data
5. Maximum data
6. Multi-line
7. Multi-page
8. Excess data
9. No data
10. Extra invalid records
11. Constructed records

Input

1. Valid input
2. Invalid input
3. No input
4. Maximum input
5. Default values
6. Hardware conditions
7. Duplicate control card (batch)

Execution

1. Terminal variations (on-line)
2. Execute again
3. Execute using special processing
4. Forced during execution:
   - unit not available
   - unit quits in middle of output
   - unit quits during handshake
   - hardware paths
   - execute restart capability
   - buffer size variations

Result verification

1. Check output conforms to standards
2. Check data usage via other software
3. Verify that invalid data is rejected
4. Verify addition via tools
5. Verify data set changes via tools
6. Verify delete via tools
7. Verify that only requested data is present
8. Check multi-page output
9. Check multi-line output
10. Check that receiving system accepts data
11. Check that receiving system rejects invalid data
12. Check messages received
13. Verify new screen restrictions (on-line)
14. Verify subsequent runs use of data (batch)
15. Check run times (batch)
16. Check control totals (batch)
17. Check log information (batch)
18. Check report break-point (batch)
19. Check for $0, -, +$ values
20. Check tables for first, last entries
21. Verify sort
22. Verify data manipulations
23. Verify translations

Checklist construction

A checklist is used to tie together the testable items, the test area, and the software functions developed in the previous sections. Appendix B contains a detailed checklist for each software function and a one-page summary of all checklists. Testers can use the checklists to formulate the contents of their testcases.

Using the checklists in appendix B is simple. For a software function, just examine the testable items in each test area and determine if the testable item should be included. Include from the checklist only those items to be tested. The only criteria for excluding an item is that the item does not apply, for example, the update software does not require tests that address hardware conditions. Rather than discussing the reasoning behind including or excluding any specific item from a checklist, develop "prototype" checklists. These checklists are tester-oriented. The testable items identified are the "most likely" candidates for each software function. The tester should select those items required for verification of the specific software being tested. Figure 5-1 is the start of the checklist for add software.

ADD Software Checklist

1. Data set conditions:
    a.   valid records
    b.   invalid records
    c.   every field supplied
    d.   minimum data
    e.   excess data
    f.   no data
    g.   constructed records

Fig. 5-1. Partial checklist for ADD software.

An initial use of these checklists is to manually check that current test procedures address the suggested items. This might lead to expanding testable items and expanding the procedures.

The checklist summary shown in FIG. 5-2 also contains a priority associated with each testable item (H = high, M = medium, and L = low). This priority is a relative indicator of the importance of the testable item for that software function. Obviously, the more critical software areas must be tested more thoroughly. Prioritizing provides a method of choosing which testable items to emphasize during testing. The priority could, and should, be changed as software is re-released with modified functions. Change would be caused by the tester's review of the modified software and the increased or decreased importance of testable items in testing the new versions.

Levels of test coverage

Test coverage is divided into levels similar to the testing levels discussed earlier. The coverage levels are based on what is being covered and are organized so that the nth level of test coverage provides a lesser degree of system cover-

Checklist Summary

Software Function

T E S T	A R E A	Testable Items	A D D	C O M M	D E L E	I N T F	Q U R Y	S T D P	U P D T
D A T A	C O N D	Valid records	H	H	H	H	H		H
S E T	I T I O N S	Invalid records	H	H	H	H	H		H
		Every field supplied	H		M	M	H		H
		Minimum data	M		M		H		H
		Maximum data				H	H		H
		Multi-line		H			H		H
		Multi-page		H		H	H		H
		Excess data	M	M		M			
		No data	L	M		L	L		M
		Extra invalid records					H		
		Constructed records	M			H			
I N P U T		Valid input	H	H	H	H	H	H	H
		Invalid input	H	H	H	H	H	M	H
		No input	M	M	L	M	M	M	M
		Maximum input	H		H		H	H	H
		Default values		H	M	M	M	L	M
		Hardware conditions		H					
		Duplicate control card	H		H	H	H		H
E X E C U T E		Terminal variations	H	H	H	H	H	H	H
		Execute again	H	H	H	H	H	H	H
		Execute using special processing	H	H	H	H	H		H
		Hardware variations		H					
		Hardware paths		H		H			
		Execute restart capability	H	H	H	H	H		H
		Buffer size variation	H	H		H			

Fig. 5-2. Checklist summary.

Checklist Summary

Testable Items	ADD	COMM	DELETE	INTF	QURY	STDP	UPDT
R **V** Check output conforms to standards	H	H	H	H	H	H	H
E **E** Check data usage via other software	H		M		H		H
S **R** Verify invalid data rejected	H	H	H		H	M	H
U **I** Verify addition via tools	H						
L **F** Verify change via tools		H		H			H
T **I** Verify delete via tools			H				
C Verify only requested data present	H			H	H		
A Check multi-page output		M	M		M		M
T Check multi-line output					M		
I Check receiving system accepts data				H			
O Check receiving system rejects invalid data				H			
N Check messages received		H					
Verify new screen restrictions						H	
Verify subsequent runs use of data	H	H	M	H			H
Check run times	M	M	M	M	M		M
Check control totals	M	M	M	M	M		M
Check log information	L	L	M	L	M		L
Check report break-points					M		
Check for 0, −, + values					H		H
Check tables 1st, last entries					H		H
Verify sort				H	H		H
Verify data manipulations				H	H		H
Verify translations				H	H		H

Software Function column headers: ADD, COMM, DELETE, INTF, QURY, STDP, UPDT

Left margin labels: RESULT / VERIFICATION

Fig. 5-2. Continued.

age the n + 1st level. Before describing levels, however, it is necessary to define some terms:

Element The concept of module testing is not adequate for complex software systems because these systems are composed of more than modules. The term *element* refers to the individual structural entities of a system, such as screens, transactions, databases, etc.

Linkage A combination of two or more related elements is a linkage. An example is the linkage of a program (one element) with the databases (one or more elements) that the program uses.

Function Used in this section with the connotation of a testable function, not an operational function. Software is specified by operational functions such as "add a user." The process of converting operational functions into testable functions is an analysis called decomposition. For the example of adding a user, let's assume that analysis determined that:

- The user should have a valid ID.
- Adding the same ID twice will not be allowed.
- The user ID will be added to a table that specifies output destinations.

These three functions would be considered decomposed testable functions.

Item Testing a software function is done by executing one or more testcases. Testcases execute the software to determine the response to selected input. The input is selected to exercise various test conditions that, in this book, are referred to as "items to be tested" or "testable items." Just as software operational functions are decomposed into testable functions, the testing of a function is decomposed into the testing of testable items. This terminology was used previously to describe the entries in the checklists developed for testing.

Defining the levels

Element coverage Coverage at this level assures that the testcases exercise the software system elements—transactions, batch runs, databases, and formats. A testcase with no data could be considered to have covered a transaction, such as when issuing a call and verifying that the expected error message is returned. The analogous system testing level is basic integrity testing that ensures the existence of the software that supports various elements (e.g., transactions, formats, etc.).

Linkage coverage   Coverage at this level assures that the testcases exercise linkages between related elements:

- The databases accessed by a transaction, batch run, or format.
- Subsequent or background transactions initiated by the initial transaction (one special case is a dynamic interface to another system).
- Action/verification pairings such as create, then find and verify or update, then find and verify, etc.

Testcases for linkages must be specifically designed and must use carefully selected data to ensure that the linkages are exercised. The comparable system testing level is the format level, which assures that one instance of each software function is executed.

Function coverage   Coverage at this level assures that the testcases exercise all variations of software functionality. This level of coverage ensures that testcases verify algorithms. To economically verify algorithms, testers must use their knowledge of the software to eliminate redundant testcases. The analogous test level is function testing that verifies correct operation of complex software capabilities.

Item coverage   Coverage at this level assures that the testcases verify all the testable items as described earlier in the chapter. This level is not comparable to workflow testing but does provide a greater degree of coverage than the function coverage level.

Document use

Documents are part of the system from the user's viewpoint. The user expects documents, such as manuals that correctly depict the system's operation. These documents form a basis for testing. Testing ensures that the documents describe the system and that the system works as described in the documents. Examples and statements about system operation found in user documentation should be viewed as requirements. At least one testcase should address the contents of each document and it should be changed when the document is changed.

Testcase life span

Testcases are designed and developed to be used in a system's regression test package for the next release. A regression testcase assures that the functionality in old releases still works correctly and supports the user's needs. Testcase development is not a throwaway activity. Testcases that are developed early in the software development cycle should be transferred to the next phase. The testcase might need changes, but the same testcase can pass from requirements to design to development to product test to acceptance test.

A testcase might not be needed in the regression test package if it is redundant but that decision is usually made after testing is completed. A testcase that is in the regression test package can become obsolete because the software functionality changed. A testcase can also become obsolete because the functionality it tests has remained unchanged and fewer testcases are necessary to ensure that the system still operates as expected.

Summary

A good methodology enables users to begin doing the test job without much training. The testing methodology presented in this chapter is based on five truisms:

1. Any test is better than no test.
2. The ultimate test never works.
3. Test where the problems are.
4. Use black-box techniques for system testing.
5. Automation pays.

If a novice tester applies only these truisms to the job of system testing, the result would be automated testcases that find problems, supply reasonable coverage, and are not too complex.

Testers are concerned with the questions "What to test?" and "How well was it tested?" By characterizing software by the functions it performs, defining test areas and testable items, and developing a test checklist for each software function, a structure can be formulated that answers both questions. Any testing organization or individual can customize the checklists to meet their specific requirements. The methodology is extended to address the role of automation, how documents relate to testcases, how to determine coverage, and how to test throughout the life span of a system.

6

Controlling
the test process

How do we ensure that we don't get into an endless loop of test, fix, improve, retest? How can we limit unexpected software changes especially? One easy way is to divide the test process and the test period into phases: start-up, high-discrepancy identification, and final. The characteristics and activities associated with each of these phases is discussed below. This chapter concentrates on controlling the test process during the time the testcases are actually being run.

IAM test environment

This section describes a hypothetical test environment for the IAM system so the control process can be discussed relative to a known environment. Many tools will be discussed in this section and in later sections that are part of this environment. Chapter 12 provides a fuller description of these tools.

A major IAM system release is produced annually to introduce new features or major revisions. Minor releases are produced bimonthly. Test support must be available for up to two major release levels in the field at any point in time. Also, the software for the next major release is being tested concurrently. Therefore, the testing organization will have at least three major levels of the system concurrently under test.

A matter of great concern in this environment is the release content. In this test environment, the planned changes to the system are tracked via SMART (Software Maintenance Request Tracking system). Software Maintenance Requests (SMRs) are initiated to address both internal and field-change requests through Maintenance Software Modification Requests (MSMRs) for changes to existing functions and Enhancement Software Modification Requests (ESMRs) for new features. SMART assigns these SMRs to work groups for investigation. After the investigation, SMART schedules these SMRs for specific system releases. When an SMR is resolved, the resolution and scheduling information are combined with the SMR to form a Software Modification Request Resolution (SMRR). In this book, SMRs will be the only acronym used to describe how changes are documented. SMRs contain much information, including a description of the problem, a description of the solution, and the effect on other parts of the system, including documentation.

The planned software changes and any unplanned software changes that are introduced as a result of finding problems during the testing interval are tracked via SIR (Status of Incident Reports). SIR is a test-oriented tool that tracks the testing requirements of the SMRs in SMART by release. To illustrate how SIR is used, consider a new batch procedure that minimally requires: application software, a runbook (that describes how to execute the batch procedure), and JCL (the IBM job-control language that controls job execution). For this new procedure, there are three separate items to test, and SIR tracks each item independently because each might have different problems and testing might be completed at different times. SIR also tracks internal changes identified by testers and reduces the paperwork and scheduling needed by SMRs during the system test interval.

The test organization acts as the first user of the system and, therefore, has a separate environment in which to test. This includes a separate set of databases used to test database initialization and database conversion and a separate set of program libraries produced by a build process. Developers turn over software source code (new or changed), with an associated SMR, to the Change Management Organization (CMO). The CMO uses a system called BUSY (BUild SYstem) that performs the complex process of reassembling or recompiling the software. This process ensures a consistent set of programs for a release because it propagates changes throughout the system, e.g., when a macro changes, all programs using the macro are reassembled. The CMO also uses the same source control system as the developers. This system, SOCS (SOftware Control System), maintains a library of source code that can be used to reproduce any level of software that was released.

The CMO performs consistency audits to check that software has been turned over for all SMRs and that no version of a program, macro, subroutine, etc. was turned over at an incorrect level. The audits also check that testers use only software that has been processed by the CMO, just like the customers, because another of the CMOs responsibilities is to ship the software to the customers. The CMO serves as a controlling mechanism to isolate and stabilize the testing environment as opposed to the development environment. This separation enables developers to generate changes without affecting the testers ability to continue to test.

Testing a very big system must be done using a mechanized process that submits testcases to a test driver. The driver is a tool that executes a testcase by simulating a terminal interacting with the system and checks that the results are as expected. Whenever a testcase produces an unexpected result, that result is analyzed. If the software is not operating as specified in the documentation or a standard has not been followed, the unexpected result is then considered to be a discrepancy. To clear this discrepancy, either the software or the documentation must be changed. Discrepancies are identified via Incident Reports (IRs), and most of the information in the IR is generated automatically by SIR when an SMR item fails. IRs can also be generated to identify a discrepancy not directly associated with a known SMR. Other possible causes for unexpected results, such as the testcase or environment needing modification, are addressed by internal test organization procedures, and because the software does not need modification, IRs are not generated. Each IR contains a description and a priority and can be accessed by all project members, especially the developers. The priorities are:

C Critical—the problem is blocking further testing and must be corrected as soon as possible.
H High—the problem will have high-user impact and must be corrected.
M Medium—the problem will be noticed by the customer and should be corrected.
L Low—the problem will not impact the customer and should be corrected when a higher priority problem is identified in the same area or as human resources become available.

So far, this section has discussed the problem-tracking system, the software-build system, and the test environment. Another environmental area of concern is the testcase development and testcase execution environment. One form of automated testing emulates testers using the software, i.e., programmed testcases, often called scripts, that simulate humans interacting with the system using display terminals. There are three other ways of creating automated testcases including:

1. Automated testcase generation—using the requirements in some form to generate testcases.
2. Typical transaction stream—using the transaction stream captured during execution of normal, daily work using a prior level of the system to verify that the stream is processed correctly by the new level of the software.
3. Automated data generation—selecting a sample of possible data conditions and, using an automated execution tool, submitting the data and verifying that the results are as expected.

The automated testing discussed here assumes scripts as testcases. The control process described will be applicable to all types of automated testing, however. These scripts, which are input to an emulator, are also a type of program and, therefore, must be tested. This presents the dilemma of a possible endless chain of required testing. Because of this, testing must be intelligently controlled and evaluated.

Test process control

Controlling the test process ensures that the testing process is effective, covers all the software, and does not get into an endless loop of test, fix/improve, and retest. The mechanisms used to control the test process addresses the following areas:

- Determining what to test.
- Documenting the test results and status to determine when testing is complete.
- Evaluating the testing that was conducted.

In chapter 4, planning for testing was discussed in detail and the test plans were described. The test plan is essential to controlling testing and includes the:

- Release content by identifying new features.
- Initial tester assignments to cover the new features.
- Estimated number of scripts needing modification because of expected software changes and, if necessary, environment issues and strategy.

Also, all the SMRs scheduled for the release are reviewed to determine the expected number of lines of code and to determine the expected number of IRs. The testing period can be divided into three phases (see FIG. 6-1):

1. Start-up
2. High-discrepancy identification
3. Final testing

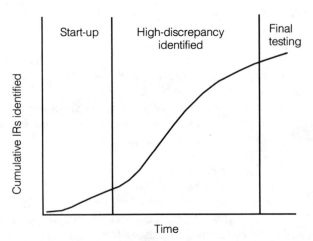

Fig. 6-1. Cumulative IRs identified versus time.

The characteristics and activities of each of these phases is discussed in the following sections. Figure 6-1 shows the number of IRs identified versus time in system test. Historically, the software IR identification has followed this pattern and many writers [MYER79 HETZ84 BIEZ84] have documented similar experiences.

The start-up phase

The start-up phase is composed of two separate tasks: the creation of the environment and identifying and resolving problems that block significant portions of the testing. The subtasks for creating the environment are:

1. Create new databases that are required to support new features.
2. Update existing databases via batch runs that reformat or add information.
3. Unload and reload databases to conform to new database descriptors (DBDs).

At this time, the operating system must be reconfigured to contain the transactions and DBDs needed to support the application software. The database activities, both creation and conversion, must be completed and verified before further testing can begin. NOTE: these activities themselves must be tested to verify that the software acts as expected (e.g., the new data appears in the correct format in the correct place) and that the installation instructions are correct. Yes, even the installation instructions must be tested. This is one of the initial tasks for the test organization.

The first type of on-line testing is Basic Integrity Testing (BIT). The purpose of BIT is to verify the assumption that the software is ready to test. BIT testcases are typically manual. Automated BIT testcases would require updating for each release and the update would take more time than executing the tests manually. BIT is conducted once per major or minor release and should generally take 8 to 16 hours to complete. The BIT status is reported using a chart like the one shown in FIG. 6-2.

Format	Program Function Keys											Commands
	1	2	3	4	5	6	7	8	9	10	11	
EQPH	–	–	0	–	0	–	–	–	–	–	–	create, verify, locate, update
EQPA	–	–	–	–	–	–	–	–	–	–	–	
EFRM	–	–	–	–	na	–	na	na	–	–	–	

The indicators are: (ok), 0 (failed), and na (not applicable).

Fig. 6-2. Sample of a BIT chart.

Each of the 0s in FIG. 6-2 are reported as an IR. The testcases are very basic. Some typical examples are:

1. Bring up a screen and press the FIND key (PFK1) without entering any data.
2. Bring up a screen, enter a known data item and hit the FIND key.
3. Bring up a screen, add a valid item, and find the item on the same screen.

The object of these testcases is to determine if the system is ready for product test. After executing all BIT testcases, the results are reviewed and, if a high enough percentage, about 95 percent, execute without error, the system is deemed ready for product test. If, however, the percentage is less than 95 percent, the start of product test should be delayed.

There are, however, some critical functions that must work before meaningful testing can begin. For the IAM system, some of the critical functions are: add a new order, process an order, ship an order, cancel an order, enter a payment, issue a credit, add an employee, change an employee's job, add new inventory, and run a trial balance. If any of these critical functions are not operational, there is a large impact on the ability to further test and product test should be delayed regardless of the BIT results.

The high-discrepancy identification phase

When a system achieves a relatively stable state—all the critical functions work, all the initial problems blocking significant testing have been passed, and all the expected software is present—the high-discrepancy identification phase begins. At this point, the system can be tested using all approaches:

1. Automated regression testing.
2. New feature testing.
3. Maintenance-request resolution testing.

Automated regression testing

Automated regression testing verifies that a system operates as it did in prior releases and has not regressed. If a capability that was supported in a previous releases is inadvertently not provided, this omission will have a large impact on the customer. Therefore, finding these cases is the highest testing priority.

The regression test package for a big system is typically composed of thousands, perhaps tens of thousands, of testcases that emulate the actions of a person interacting with the system via a terminal. A testcase performs terminal actions such as: enter data in a specific field, depress a program function key, check the message returned, and/or check a specific returned data field.

Because a testcase, in our environment, is a program, it can also do operations such as if-then-else, do while, comparisons, or read a file.

A cornerstone of the regression test package is a series of workflow scripts that verify that the recommended standard workflows, used by the system's users in the field, execute as expected. These scripts must be updated during the test period if a new capability changes the workflow.

Tools are needed in the hypothetical IAM test environment to track testcase status and execution. With many testcases and many releases being tested concurrently, a critical factor for adequate control is to maintain an approved list of applicable testcases. A testcase's status is maintained in the Testcase Activity Report (TAR), a test control system that contains:

- A testcase identifier.
- The release for which the testcase was first applicable.
- Whether the testcase is manual or automated.
- Whether the testcase development is completed, in planning, or in progress.
- The current status—working, error-blocking completion, or not yet run.

This report is the source of testcase information and serves to monitor new testcase development and to provide the list of testcases used in the testcase selection process.

Testcases that will be contained in the regression test package for a given release are determined by selecting testcases from TAR. Two modes of selecting and executing testcases are available—Manual Queuing (MQ) and SeLection and OPtimized Execution with Reports (SLOPER). MQ operates in a single-user mode that allows a tester to queue up a single testcase or a series of testcases to an already-running, simulated terminal(s). The testcases then execute serially as resources become available. SLOPER operates in a regression mode that allows the tester to select all or some subset of the testcases based on application, subsystem, or similar criteria. After testcases are selected, SLOPER apportions the testcases across several simulated terminals to ensure an even workload.

All testcase executions are recorded and summary reports are generated. The most important report is the error report that is generated whenever a testcase has errors. This report is used by the testers to determine the cause of the error. Testcase status reports show testcases run, testcases with errors, testcases without errors, and testcases not run. These reports are available for each tester or for groups of testers.

The progress of regression testing is measured in terms of testcases executed without error (see FIG. 6-3). The sharp dip at 45 is caused by reinitializing all the scripts to a status of not-yet run and running them again during the final phase of testing. Each testcase with an error must be reviewed and either

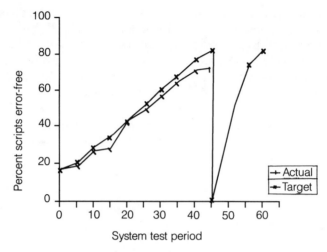

Fig. 6-3. Regression status.

the testcase is modified because the system changed as expected i.e., the system was used to debug the testcase, or an IR is opened.

New feature testing

The purpose of new feature testing is to verify that a new capability introduced into the system operates as documented and conforms to the software standards. Testing new features is documented in the feature test plan. This plan is reviewed by testers, developers, and requirements personnel before the start of testing. The testcases identified in the test plan are designed using a level structure to provide control and speed up IR identification. The initial level is the format level that checks all the transactions initiated by an on-line format. The next level is function testing, which checks data variations of transactions that are algorithmic in nature. The final level is flow testing, which checks the interaction of related transactions.

The progress of new feature testing is measured by the percentage of the test plan completed. A graphical display of percentage complete versus time in test is developed and included in the weekly status report (see FIG. 6-4).

While it is more effective to develop and use automated tests for new feature testing, manual testing is typical because developing automated tests requires a long lead time and the feature is undergoing functional changes during test, only reaching final form at the end the testing period. Whenever possible, extra resources are assigned to automate tests during the test period. Automated tests are designed to become part of the regression test package for future releases.

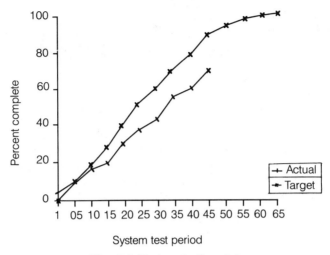

Fig. 6-4. Feature testing status.

Software modification testing

Manual or automated testing of Software Modification Requests (SMRs) verifies that the modification operates as documented and that the modification has not adversely affected any related capabilities. Some of these tests are conducted by modifying existing automated testcases. Others are conducted manually and then automated for the next release. The measure of progress of SMR testing is the number of closed SMR items (see FIG. 6-5). Remember that

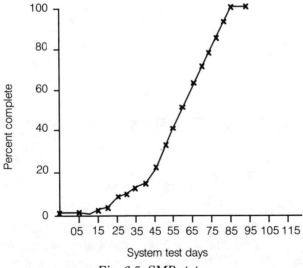

Fig. 6-5. SMR status.

the SMRs are broken down into SMR items so they can be adequately tracked and controlled. A graphical display of SMRs in the release and SMRs closed versus time is included in the weekly status report. The failure of an SMR item results in an IR.

The final phase

The final phase of testing occurs when all the software expected to correct IRs, both planned and delivered, is in the system. During this phase, discrepancies take longer to find as FIG. 6-6 shows. This graph is also part of the weekly status report. When IR identification falls below some predetermined rate (e.g., two a week), the economic view is that keeping the system in test longer is more expensive than the cost of correcting the discrepancy in a

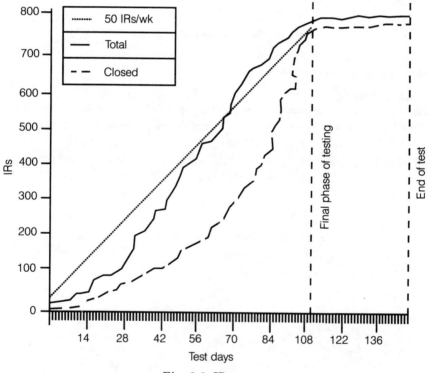

Fig. 6-6. IR status.

future release. This is true, even considering that the cost ratio of correcting a bug during test to correcting a bug in the field is 20:1 [CRAI86].

During the final testing phase before the system is shipped, evaluation testing is done. Evaluation testing ensures that the system meets the user's needs. One task is to review each discrepancy that was not corrected. The focus of the review is to determine the impact on the customer of having the discrepancy present in the software. If the impact is low, the correction is deferred to a future release. If the impact on the customer is high and the impact on the system is low, the correction is made in the current system. If the impact is high, but the risk of impacting the system is also high, management must evaluate the situation. Correcting the problem could delay the shipment of the release.

All deferred discrepancies are identified for the customer in the release package, which serves as a transmittal document and describes the release. Customers can then determine the impact of each deferred discrepancy on their own operations and submit software modification requests for discrepancies they want corrected.

Status reports

During all phases of testing, test status reports are issued daily and weekly to project managers who control the project. Each day, a one-page report is issued that contains the SMR status, IR status, testcase status, and overall quality estimators (see FIG. 6-10) [LIND88]. Each week, a report is issued containing the graphs previously identified and graphs that show a historical comparison to prior releases (see FIGS. 6-7, 6-8, and 6-9). These reports form a core of expected status information but as other interesting metrics are developed, they are included in the status package.

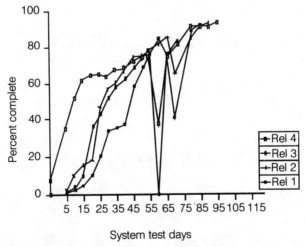

Fig. 6-7. Test status history—regression comparisons.

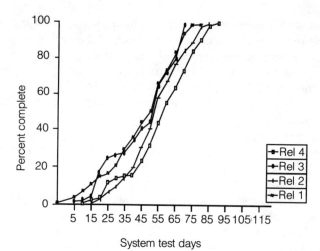

Fig. 6-8. Test status history—SMR comparisons.

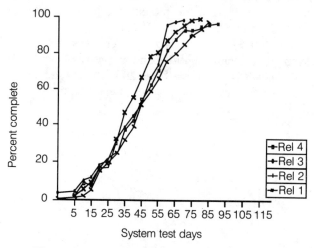

Fig. 6-9. Test status history—IR comparisons.

Test process evaluation and review

There are two aspects to test process evaluation:

1. Was the testing effective? Did the testcases cover the spectrum of what was to be tested?
2. Was the process effective? Was the strategy correct?

Testcase coverage is a complex issue for a mature system. Current test methodologies rely on requirements-based testcase design to address cover-

Section 1—SMR Status

TOTAL SMRs	=	Total number of SMR items in the release.
CLOSED SMRs	=	Total number of SMR items in STP, CAN, DEF, ROL, or STR status that indicates no more testing is needed.
SMRs in DEV/STF	=	Number of SMR items that have not yet been completed by development or that have failed system test. Additional software is expected.
SMRs in ST	=	Number of SMRs in ST that are ready to be tested.
SMRs in STI	=	Number of SMRs in STI that have been initially tested but more testing is required.
OUTSTANDING SMRs	=	Total of previous 3 categories.
DOCUMENTATION SMRs	=	Number of SMR items that represent documentation changes.

Section 2—IR Status

TOTAL IRs	=	Total number of currently identified IRs.
CLOSED IRs	=	Number of successfully retested IRs.
OPEN IRs	=	Total number minus closed.
DEV STATUS	=	Number of IRs in development with modification expected.
CRITICAL	=	Number of critical IRs that impact testing progress and should be fixed rapidly.

Section 3—Testcase Status

TOTAL	=	Total number of testcases for this release.
NOT YET RUN	=	Number of testcases not yet run.
TOTAL RUN	=	Number of testcases run.
TOTAL NO ERROR	=	Number of testcases run with no errors.
TOTAL WITH ERRORS	=	Number of testcases run with errors.
PERCENTAGE TESTCASES RUN	=	(TOTAL RUN/TOTAL) × 100.
PERCENTAGE RUN OK	=	(TOTAL NO ERROR/TOTAL RUN) × 100.

Section 4—Quality Indices

REGRESSION INDEX	=	(TOTAL RUN OK/TOTAL) × 100.
SMR QUALITY INDEX	=	(CLOSED SMRs/TOTAL SMRs) × 100.
IR QUALITY INDEX	=	(CLOSED IRs/EXPECTED IRs) × 100. Expected IRs developed from experienced-based factors associated with SMR types and lines of code delivered.
RELEASE QUALITY INDEX	=	Combination of previous 3 indices with weighting factors.

Fig. 6-10. A one-page status report.

age. The testcases are designed to verify that all the requirements are met. The testers must be concerned with coverage because any uncovered area might imply incomplete requirements or excess software. Batch test coverage is concerned with procedure coverage.

Checklists have been developed to assure functional test coverage. The checklists show the areas that should be covered for each functional type of procedure (e.g., update, inquiry, etc.). These checklists are based on input from experienced testers at test design reviews over a long period. They have since been used for many years with good results. On-line test coverage is based on transactions executed and also includes on-line interfaces. The design of on-line testcases is screen-oriented. At a minimum, each command

and program function key is verified. NOTE: for a transaction-based system, hitting a program function key with a format on the screen usually generates a transaction to be executed. Therefore, a measure of test coverage is the percentage of individual transactions executed. During the final testing phase, information is gathered from operating system log tapes to develop an on-line transaction coverage report. If a transaction was not executed, a testcase is developed to test that transaction.

After testing is completed and the release is shipped, the system testing organization issues a final status report. The purpose of this report is to evaluate the test strategy and the test process used for a specific release. The final version of each of the graphs generated for the weekly status reports during the test period (i.e., percentage of new features tested, SMRs closed, IRs identified and closed) are included. Any difficulties encountered during the test period are discussed. Recommendations for improving future strategy or test process improvements are also presented. This review serves as an evaluation of the test process for the release.

The final judgment of the release's quality is based on the SMRs reported from the field, however. These SMRs are reviewed by the testers, developers, and requirements personnel to determine where the problem should have been found and why the problem was not found before the software was shipped. The testcases are expanded to include new testing conditions for those problems that could and should have been identified during the test period. Also, SMRs should be reviewed to determine trends and to develop defect prevention methods.

Between releases, the regression test package is evaluated. It is essential to minimize the number of testcases so that the time required to execute the regression testcases does not increase without maintaining control. Redundant and obsolete testcases must be eliminated.

Summary

The controlling factors for ensuring that the system test process does not get into an indefinite loop of testing, fixing, and retesting include:

1. Tight control of the system software content using tools such as SMART and SIR. This way, only documented changes are allowed into testing. This assures that testers know what to test and when.
2. Prioritizing problems found during test so that developers have a guideline as to which problems to fix first.
3. Reallocation of resources to address activities that the status reports identify as being behind planned schedules.
4. Track discrepancy identification rate to determine when to start the final phase of testing.
5. After testing is complete: reviewing the test process for coverage, reviewing the process for areas of possible improvement, evaluating the SMRs received from the field, and evaluating and improving the regression test package.

7

Testing procedures

In this chapter, the testing methodology and the testing approach described earlier will be used to develop the test procedures. Testing procedures can be divided into five areas:

1. Determining the test emphasis—old functionality, discrepancies, modifications, or new functionality.
2. Implementing/updating testcases.
3. Executing testcases and analyzing results to determine if problems exist.
4. Resolving problems and retesting changed software.
5. Reassessing test priorities then going back to #1.

Each area is discussed in the chapter as well as concrete procedures testers should use. Procedures for controlling the databases and for stress and performance testing are also described.

Test emphasis

Test emphasis must be spread across four types of testing, each of which requires a different testing procedure. Let's assume that the software has already been released and that the current system contains unchanged code, new code, and modified code. The first type of testing is directed at ensuring that unchanged old features and capabilities operate as they did previously. Because testing uncovers problems, testers must direct their efforts at a second type of testing—retesting, which is done to verify that the problems have been corrected. The third type of testing verifies that modifications to previously existing software are correct. The fourth type of testing addresses new features.

Retesting old functionality

Retesting old functionality is also called regression testing or equivalency testing. The challenge is to select an appropriate set of existing testcases. The procedure must ensure that the testcases selected provide adequate coverage of old functionality with a minimum of redundancy. It is assumed that regression testing is automated because the testcases might have to be repeated several times. The regression test package is one method of verifying that fixes did not adversely affect old capabilities. All regression testcases should be run if there is any question about what areas of the system were affected by changes or new features.

Retesting discrepancies found during testing

When the software does not perform as documented or when the software does not conform to standards, an IR is issued to track the problem. Once the software is fixed, the discrepancy retesting procedure must notify the tester to retest the fix to ensure that it resolved the problem. Fast IR turnaround by developers enables the tester to identify the next problem earlier. Therefore, when testing discrepancies, the tester's goal must be to retest them rapidly and to ensure that the next problem, in the same area, is identified quickly.

Testing modifications

The user, in the field, reports problems via an SMR (Software Modification Request). The software is then modified to fix the problem and the test procedure must verify that the problem was fixed without adverse side effects. Sometimes, the problem is fixed, but processing in another part of the system (another component subsystem) is affected and might not work correctly. If regression testing does not cover this specific area, not only must the fix be tested, but also other areas in the component subsystem that could have been affected, as well as related processing in other component subsystems.

SMRs contain a description of the problem that should enable the tester to duplicate the conditions under which the problem was identified. When the SMR contains insufficient information, the tester must determine how to duplicate the problem or if the problem can actually be duplicated. Some examples of SMRs that cannot be duplicated are:

- If the cause of the problem was in error-handling software and the error cannot be forced in the test environment.
- If the cause of the problem is volume or load related, it might not be possible to duplicate it in a limited test environment.
- If the problem is performance related, the exact system load probably cannot be created within the test environment.

Testing new functionality

The first activity for testing new features is to develop a comprehensive test plan as early as possible. After the requirements are available, testers can use them to develop an initial test plan. At this stage, planning should ensure that all requirements are tested. If the requirements identify the need for new test tools or new test techniques, it is appropriate to begin the acquisition process at this point. Planning and tester involvement continues throughout the software development phase.

Before the software is available, testers can develop test specifications for the necessary testcases. These specify the actions that will be performed and the results that are expected. Test specifications are more detailed than the test plan and specify how testcases should be written. Once test specifications are written, testcases can be automated. However, it might be premature, because the users could review the system and change the requirements. Or, after user documentation is available, it could be discovered that the testcase results are different from what was originally expected.

Implement/update testcases

Testcases are logical sequences of action(s), from a user's perspective, that are executed with the objective of finding problems in the software. Testcases contain one or more actions (transactions) and can be automated. They are also usually developed to test a requirement or a functional area. A functional area is a subset of the system's capabilities, which can be tested independently from other system capabilities. After a system has been in the field, functional areas can be added or changed to enhance the system's usability. Testcases that were originally effective might no longer be as effective and might not even work. These testcases must be updated to provide the same coverage as they did before the software changed.

The following sections discuss the procedures used to develop new testcases and to update existing ones. Testcase implementation is divided into two phases: testcase design and testcase development.

Testcase design

The following testcase design criteria are part of the procedures for testcase implementation. A testcase must:

- Find problems.
- Cover requirements and functions.
- Be repeatable and deterministic.
- Be maintainable.
- Be nonredundant.

Find problems   The objective of testing is to measure the quality of the software being tested by finding problems. The process is to execute testcases and analyze the results within time and budget constraints. An alternate design criterion to finding problems is to prove that the system works correctly. However, when testing large (500,000 lines of code) data-oriented systems, a large data sample size is needed to show that the system works. The time required to execute the number of testcases needed to supply the data sample size can only be justified, economically, for life-critical software such as atomic reactor control systems. For usual business applications, finding problems is a more practical and economical design objective. It is important to recognize that the problems testers want to find are those that the customer would find and want corrected. Fixing a problem during the test period is less expensive than fixing a problem once it has reached the field. Therefore, the data used for testing should emphasize lifelike conditions and testcases should emulate customer interactions (workflows).

Cover requirements and functions   Testcases must address the coverage of the requirements. It is expected that requirements will be testable. The development of testable requirements, although important to software testing, is not covered here. Requirements are covered when a testcase has, as its objective, to verify that each specific requirement has been satisfied. Some testcases demonstrate that several requirements are satisfied, while other requirements need several testcases to demonstrate that they are satisfied. For purposes of coverage, each requirement must be traced to at least one testcase. For purposes of testcase selection (when a requirement is changed, what testcases should be run?), all the testcases that verify the requirement must be run. The forward reference set (requirement to testcase) must contain at least one entry, while the backward reference set (testcase to requirements), must identify all the requirements verified.

Coverage functions   Coverage can be measured in terms of externally observable functions. The analysis associated with the coverage of a function consists of a decomposition of the software into functional areas and data dependencies. These relations are expressed as a matrix of possible test conditions. The testing procedures require that testcases be developed to address at least each column and each row once. Note that this analysis can also be used when no formal requirements exist. A simple example to illustrate the decomposition process is to test the function "find an order" within the fictional IAM system. Assume that the customer orders are entered with four possible key fields: internal order number (intordno), related internal order number (relintordno), customer order number (custordno), and customer identification (custid). Further, assume that there are three order types: Rush (for which the customer pays a premium, if the order is delivered on time), Normal, and Hold. The matrix that represents a decomposition of the find function is shown in FIG. 7-1. Note that this matrix assumes that the testcases will use only a single-key field as input. The matrix could be further expanded by using combinations of key fields such as input e.g., intordno and custordno.

		Field for find			
Order type	int #	Related Int #	Cust #	Cust id
R				
N				
H				

Fig. 7-1. Test coverage matrix.

Twelve testcases are required to cover the 3 × 4 matrix in FIG. 7-1 entirely. This is a simple illustration, because in practice, matrices of this sort can be on the order of 10 × 30, and require 300 testcases. Note that if combinations of input were used in the sample, the number of columns would be 16. The smallest number of testcases to ensure functional coverage is 4: find a Rush order by intordno, find a Normal order by relintordno, find a Hold order by custordno, and find a Normal order by custid. This covers the diagonal and one more case that allows the tester the option to choose the most important testcase. In this example, Normal was chosen as the order type because it represents the most often-used order type and should, therefore, be tested more extensively. The matrix of FIG. 7-1 is shown again in FIG. 7-2 with an X showing each of the testcases chosen.

		Field for find			
Order type	Int #	Related int #	Cust #	Cust id
R	X			
N		X		X
H			X	

Fig. 7-2. Minimum test coverage matrix.

Consider now how a tester might increase the minimal set of testcases for better coverage. Because we know that Rush orders generate additional revenue, if they are delivered on time, we could expand the number of testcases by adding R with relintordno, R with custordno, and R with custid. Note that the testcase N with custid can be removed because the testcase R with custid provides coverage for custid. While it was not specifically mentioned, testers know that H orders are stored in a separate file called the hold-order file. The testers know this because they test the initialization and maintenance routines for the hold-order file. An additional testcase H with intordno could be added to increase the data coverage. The matrix of FIG. 7-2 is shown again in FIG. 7-3 with a Y showing each of the additional testcases chosen.

By careful planning, the number of testcases was reduced from 12 to 7 and still provided functional coverage because a testcase addresses each row and each column. For a 10 × 30 matrix requiring 300 testcases, diagonalization

Order type	Field for find			
	int #	Related int #	Cust #	Cust id
R	X	Y	Y	Y
N		X		
H	Y		X	

Fig. 7-3. Expanded test coverage matrix.

would start with 30 testcases and then perhaps double that to address economically important functions or data flow. This would reduce the number of testcases from 300 to perhaps 60, with each row and each column having at least one testcase.

Repeatable and deterministic   These are distinct design criteria but they are closely related. A testcase is repeatable if it can be rerun without any intervention. A testcase is deterministic if the output can be specified before the testcase is executed. If the testcase leaves the database unaltered, then it can be both repeatable and deterministic. Testcases that don't change, are databases are usually designed to establish software readiness rather than to test functionality and are useful in establishing sanity. But, the more effective (find problems) testcases, alter the database during execution. Special effort is required to make testcases that alter the database both repeatable and deterministic.

For example, consider a testcase that assigns the next free inventory number to an order. NOTE: for most inventory systems, once the inventory number is assigned, it cannot be unassigned, even if the order is canceled, because the cancellation must also be tracked. Therefore, each time the testcase is executed, a different number is assigned. Thus, the testcase is not completely deterministic. There are two ways to make testcases deterministic.

The first way is for the testcase to reset the databases to the expected initial condition as the last step of its execution. This is the best approach because the testcase will be both deterministic and repeatable, provided it cleans up after itself. This is not always possible, however.

The second way is to externally refresh the database to the original condition before rerunning the testcase. Refreshing and then rerunning the testcase will result in identical assignments each time unless the sequence of testcase execution causes differences. However, because refresh is an intervention, the testcase is not repeatable. The test procedures must specify how often to refresh the databases. If the testcases are deterministic, but not repeatable, the databases must be refreshed frequently. If the testcases are repeatable, but not deterministic, the databases can be refreshed less frequently. For systems with large databases, the time needed to refresh the databases could be long e.g., hours. To refresh the databases to assure testcase repeatability, usually requires extra resources such as copies of the databases or second-shift support to refresh off-hours. The cost of supporting refresh efforts might make this method uneconomical.

Maintainable The same design principles that ensure that program code is maintainable apply to the procedures for writing maintainable automated testcases. Other maintainability design principles are:

Keep going, if possible Testcases are designed to find errors but if the testcase stops processing with the first error, other possible errors will be masked. There are circumstances that render further processing a waste of time, however, because all further activities will result in errors relating to the first error. Except in these cases, case processing should continue. Testcases that stop for each error require more maintenance because they require some change for each error discovered. Therefore, to minimize the number of times that a testcase must be changed, write testcases that report the largest possible number of errors in any execution.

Program defensively Good programming practices recommend that programs be designed so that errors in other programs do not cause problems in the current program. To accomplish this, programs should validate input before processing it. In testcase design, the object is to prevent system errors (or authorized changes, because the change can represent a departure from the expected action) from causing problems in the testcase. The technique to prevent this is to check all results before using them. By designing testcases this way, problems are identified at the point of origin and there is no need to retrace the actions that caused the problem.

For example, using IAM's order system, is a testcase that consists of updating an order. Assume that the testcase adds an order, updates it, and then checks that the update worked by finding the order and checking the updated field. If the add did not work, the first error indication might be that the expected update has not occurred. The tester would then have to trace the testcase actions back to the failed add. It is far easier to maintain the testcase if a check is made to see that the add order transaction completed correctly. If it did not, the testcase should report the failure and proceed if possible.

Verify simple before complex Structure testcases to verify simple operations before complex operations. This builds confidence that the software operates correctly in an incremental fashion. It is far better to determine that many simple operations work correctly before finding a complex operation that does not work. Also, if the testcase does not adhere to **Program defensively**, the failure of the complex operation might limit the ability of the testcase to exercise later operations. Knowing that simple operations work makes it easier to pinpoint an error and to maintain the testcase.

Follow the flow of the software Even using black-box techniques, the testcase should follow the sequence of the software. The testcase should provide input in the order that the software processes input. This minimizes the possibility of masking a problem. For example, a program uses four input values, which are checked in the order 1, 4, 3, 2 (i.e., if no input is provided, the software reports an error about input 1 missing. After input 1 is provided, the software reports an error about input 4, etc.). Provide the input in the order of errors being reported rather than in the order 1, then 1 & 2, then 1 & 2 & 3, etc. A problem of incorrectly accepting input in the order 1, 4, 2 might be overlooked if the program's sequence is not followed. If the software changes the order of the execution, the testcase will report errors that indicate the change. This order of testcase development later assists the tester in identifying changes that might impact other testcases and speeds up maintenance because the cause is known and the impact can be determined before executing other testcases.

Make repeatability "industrial strength"   The repeatability of a testcase is a critical design criteria. It is not enough to design for repeatability, it is necessary to overdesign for it. The usefulness of a repeatable testcase over a nonrepeatable testcase is great. Environmental problems can invalidate the first execution of a testcase or testcase coding errors can exist. If the tester must intervene before the next execution of the testcase, the productivity of the tester is severely impacted.

A testcase, even though it changes the database, can be made repeatable by resetting the database at the end of the testcase, thus undoing the changes. But this is not enough to make the testcase "industrial strength." The testcase must first initialize, then run, then reinitialize. The first, initialize (part of the set-up), addresses the case of the testcase not completing previously and, therefore, not executing the cleanup (the reinitialize). This initialization prepares the database and might include adding, changing, or deleting database items. It can be streamlined to not check results but to just execute the necessary interactions if the simple tasks have been tested earlier in a less complex testcase. The second reinitialization is a cleanup and should check that the adds, changes, or deletions worked properly and leave the database in the expected condition.

Much consideration should be given to overensuring repeatability to increase the robustness (usefulness) of a testcase. When reviewing a testcase, the question, "Would the testcase be repeatable if it stopped here?" should be asked and answered frequently. Remember, an executing testcase could be terminated at any given point because of outside influences, such as test driver problems. The more robust a testcase, the less maintenance it will require.

Assume that the results will change   Over the life span of a software system, testcase results can change. New requirements or maintenance can result in changes. Therefore, a testcase verifying the software's operation will have different outcomes than it previously did. If the testcase is validating results (as it should, see **Program defensively**) then the testcase must be changed. To ease testcase maintenance, make the checking of results "easily tunable."

For example, if a testcase checks for messages returned by the software, keep the messages in a separate file. Then when the messages change, the tester can change the message file, recompile it (if necessary), and all testcases using that message will work. Another example is validating the output of a report. If the testcase checks the entire report against a previous copy using a differencing utility (to save analysis time), then the new report has to be verified and saved for the testcase to work. If, however, the testcase looks for the specific occurrence of a data item (e.g., a line with an expected value), then the testcase could work regardless of the format of the output.

Designing testcases so that valid result values can be modified reduces maintenance significantly.

Nonredundant   Each testcase should provide unique information about the software. Testcases that do not provide new information should be eliminated because executing them, analyzing results, and maintaining them, uses valuable resources. Unfortunately, it is not easy to determine that testcases are redundant. During the life span of a software system, the number of testcases increases as testcases are added to test new features and changes. Often, testers who add new testcases are not the same as those who developed the previous testcases. These new testers often design redundant testcases because they lack in-depth knowledge about previous testcases. Redundant testcases can also be generated by testers testing the same area who want to test in more detail. As the testers become more experienced, they tend to increase the number of data variations and functional combinations. By documenting the purpose of each testcase and by periodic review of coverage, redundant testcases can be minimized.

Testcase development

The development of a testcase can be separated into three areas: testcase description, testcase "coding," and the environment for testcase development. Procedures must be developed for each area.

Testcase description   Each testcase should have the following information associated with it.

- Purpose. What requirement/functionality the testcase verifies, the software release and level, and the dates the testcase was developed and modified.
- Description. The method used to achieve the purpose.
- Screens used. A list of all the screens used by the testcase. NOTE: if the software is not screen driven, then describe the functionality used.
- Database requirements. A description of any database preconditioning that the testcase requires.
- Reference documentation. A list of the documentation that was used to develop the testcase or that the testcase verifies.
- Implementation requirements. It is expected that almost all the testcases will be implemented in the same manner. There are, however, expected variations, any of which must be identified.

- Special implementation requirements. Any unexpected variations in the implementation of the testcase should be identified.
- Input. A description of the data input.
- Expected output. A description of the expected results.
- Analysis information. If the expected results have a range of acceptable values, define the method for determining if a result is acceptable.
- Environmental requirements. Supply the processor time, the CPU wall clock time, and the memory requirements needed to execute the testcase.

Testcase "coding"   Usually the term *coding* is applied to creating automated testcases. But, for a manually executed testcase to be repeatable, a written sequence of actions must be followed. This sequence is an English language codification. Thus, the term *coding* is equally applicable to a manually executed testcase. While the code generated is dependent on the language and the test driver, the testcase developer should follow good coding practices. When coding a testcase, follow the KISS (Keep It Simple Stupid) principle. Testcases require maintenance, and if they are easy to read, the maintenance effort is reduced. Avoid subroutines and/or functions. Comment liberally. In general, follow the same guidelines to generate a testcase that are followed to generate application code.

Testcase development environment   The testcase development environment is just as critical as the application development environment. Automated testcases should be maintained in a configuration control system. While this might be a luxury for a small, nonvolatile system, it is essential if multiple releases of the system are being supported. As the testcases are changed to address software changes, the old versions of the testcases must be kept available to test the systems in the field and the new versions must be available to test the system being developed. If problems are found in the field, testers must be able to recreate the problems in the old version of the test system to ensure that the corrected software resolves the problem.

Each release of the system must have separate, physical databases because the databases change with each release. Also, each tester must have a data environment that is separate from other testers for each release of the system. This can be done by identifying the data used by a testcase and getting an agreement that no other tester will use that data. Separate data can be guaranteed by separate instances of the data. Also, a single tester testing multiple functions should use different data so that all testcases can be run in parallel without interfering with each other. This can be done by maintaining individual tapes, individual databases (or subsets of databases). Another, not-as-secure method, is to partition the database using some key factor.

For example, when testing inventory-related processing, the tester would use the prefix "INV" on all orders, but when testing order processing, the tester could use the prefix "ORD" on all orders. This method can lead to some surprises. For instance, when testing the orders due tomorrow, the tester may unexpectedly find some INV orders. The testcase should expect extraneous orders and ignore them.

Testcase update

The impact of software changes in response to user requests or for improvement is that testcases must be updated. Updating is necessary whether the testcases are automated or manual. Because a testcase mimics a person using a terminal or reviewing a batch report, any change in input or output that the testcase might use requires a change to the testcase. Some common changes and their impact are:

- The system screen that starts an interaction is changed to provide additional functionality. All testcases using the screen must be updated. The impact depends on how the testcase is coded, which depends on the testcase driver. For example, if the screen (or some symbolic representation of the screen) is included within the testcase, it might only be necessary to "recompile" the testcase. If the testcase uses "tabs" to position the cursor, additional coding could be required. If the testcase uses row and column values to identify fields for data entry or for verification, changes might be necessary. If the testcase is a record-replay type, it might be necessary to use a screen update tool to update the testcase.

- The position of the cursor, either at presentation time or after a return, is often changed, usually to eliminate unnecessary key strokes required to position the cursor to the next field. The standard is that the cursor should start at the field where an operator would first enter data and should return to the field where an operator would type data if the interaction was successful. Testcases that assume cursor position must have code added that repositions the cursor, if the assumed position changes.

- Messages are changed, usually to improve readability. If messages have been collected in a directory used by the testcases, then the directory is updated and the testcases are automatically modified. If the messages are coded as an integral part of the testcase, then the testcases referencing the messages must be changed to reference the new messages. If the old message can be overwritten, then it might only be necessary to recompile the testcase. Again, remember that old versions of the testcases must be available for regression testing. Keep in mind that there are interpretive test drivers that would not require recompilation.

All testcase maintenance activities should be tracked. Because this activity is time-consuming, it must be monitored to determine if extra resources should be devoted to testcase updating.

Executing testcases and analyzing results

After implementing testcases, testers must execute them against the system being tested. The environment for executing testcases for large systems is discussed in further detail later in this section. The results from executing the testcases must then be analyzed. Any difference between expected and actual results must be analyzed to determine if a software problem exists.

Testcase execution environment

The environment supporting the execution of the testcases must possess the following attributes:

- It must have a means to collect the results of testcase execution. The status of each testcase (run with errors, run without errors, not yet run) executed, or planned to be executed, must be maintained for the system under test. If more than one release of the system is under test, the execution status must be maintained separately for each release. If testcases encounter unexpected results, the results must be reported, in full, via a hard copy or be saved on the system for later analysis.

- It must support the execution of a single testcase by a tester who is developing a testcase or who is rerunning a testcase that previously ran with errors.

- It must support the execution of multiple testcases in parallel to regression test the system.

- It must support several releases of the system that might be under test concurrently.

Large testing efforts require that automated tools be used to track status. Therefore, the environment must support status reporting with a minimum of effort. Major status indicators should be condensed to fit the one-page report described in chapter 6, Controlling the test process.

Problem identification

A problem occurs when there is a discrepancy between how the software works and how the documentation says it should work, when the software does not conform to standards, or when it does not meet the customer's needs. For each problem, identified by analyzing testcase results, an incident report must be issued.

The terms *problem* and *incident* are loaded terms. Problem seems to convey implicitly that a developer made a mistake. Incident seems to convey the impression that no one is responsible. Therefore, the term *discrepancy* is used for the remainder of the book to describe all software problems. During testing, discrepancies and issue incident reports (IRs) are issued. An IR should contain the following information:

identifier   A unique identifier, for a large system, the first part of the identifier is the release being tested, the second part is the area being tested, and the third part is a sequential number.

abstract   A summary identifying the discrepancy for easy reference.

description   A detailed description of the discrepancy including: input, expected output, actual output, date and time found, the environment, activities performed to isolate the discrepancy, and information needed to recreate the discrepancy, such as testcase ID.

priority   An assessment of the impact of a discrepancy on testing and on the user—*C, H, M, L*, where *C* (critical) means testing is being held up and the discrepancy must be corrected ASAP; *H* (high) means testing is impacted and the discrepancy must be corrected within three days, *M* (medium) means testing can progress but the discrepancy should be corrected within 10 days, and *L* (low) means no impact on testing and the discrepancy should be corrected when convenient.

person responsible   Person responsible for the correcting discrepancy.

status   A field for the status that is used to track the IR. Regardless of whether a procedure is manual or automated, a discrepancy and its status must be tracked.

It is the tester's responsibility to determine who is responsible for correcting any discrepancy. That job can be easy or difficult depending on the specific discrepancy. In an easy case, the tester notifies the developer, who the tester thinks is responsible for the discrepancy, and the developer accepts the responsibility and agrees that the discrepancy will be corrected. In another easy case, the software is working correctly but the document describing the operation is incorrect. The person responsible for the document is notified and accepts responsibility for making the change.

Difficult cases occur when a discrepancy is found in an interface and the developer on either side of the interface could be responsible but neither accepts responsibility. The tester must then raise this issue to management to ensure that it is addressed in a timely fashion. In cases where a discrepancy is found in one area but is caused by a discrepancy in a second area, a second "related" IR should be opened to track discrepancy resolution. When the second IR is resolved, the second discrepancy is retested, and if the discrepancy is corrected, the first IR can be retested.

Results analysis

When the results produced by a testcase execution differ from the results expected when the testcase was developed, the tester must determine if the difference is a result of a software change or a discrepancy. The tester can use documentation, standards, and the tester's knowledge of the customer's needs to decide.

Documentation   Documents that can be used to check the software include: requirements, design specifications, documents delivered to the customer (user guides, runbooks, practices, etc.), and internal memoranda. Requirements documents should have been written as assertions such as "The software will do _____." A testcase is then designed to verify, "Does the software do _____?" If the testcase description identifies what the testcase is addressing, and if the software does not operate as expected, the discrepancy can be traced back to the original requirements document. User documents usually describe how to use the software. A typical description would be "Put these values in these fields," "Perform this action," and "The results will be _____." The testcase is designed to follow the instructions and to verify that the expected results are obtained.

Standards   Standards are written and unwritten. ALL standards must be followed! One example of a written standard might be a standard specifying the Job Control Language (JCL) to be used for batch procedures. The user community and the developers establish and document a standard that defines conventions for: data set names, device types, job step restartability, and check-pointing. All the JCL developed for batch runs must then be checked against this standard for conformance. If the JCL does not conform, an IR is issued. An example of an often-unwritten standard is the use of the Program Function Keys (PFKs). While all component subsystems might use PFK1 for FIND and PFK2 for FORWARD, the use of other keys might not be as clear. This leads to the possibility of two component subsystems within a big system using the PFKs inconsistently. Testers must be aware of this possibility and enforce the unwritten standard to ensure that the PFKs are used consistently. Better yet, testers should see to it that the standard is written. An example of tester experience being used as a standard would be finding misspelled words in messages returned by the software. This might seem to be a low-priority discrepancy but experience has shown that misspellings are reported as problems. Therefore, correcting misspellings is important.

Meeting customer needs

Testers must adopt the attitude that they are the software's first user. All software must be examined with a view towards the eventual impact on the customer. While most of the customer's needs are addressed in the requirements, some of these needs are unexpressed human considerations that are well known to experienced testers. For instance, the cursor on a format should always be initially positioned at the field where data input will normally begin. Positioning the cursor at a home position requires the user to tab before entering data. These extra keystrokes are costly to the customer when viewed in their environment of thousands of transactions per day. Discrepancies such as this must be corrected immediately because the customer's needs are foremost.

Even the most experienced testers cannot foresee the impact new software will have on the customer, however, and customers should be invited to come to the development site and test the new features in the test environment. This early customer involvement affords the opportunity to correct some customer-perceived problems before the software is shipped. Sometimes, despite written requirements and demonstrations on prototype systems, the final implementation is not exactly what the customer expected. Early identification of such discrepancy enables developers to either correct the discrepancy or at least to alert customers so that their expectations match the reality of the delivered software.

Handling discrepancies and changes

During testing, software changes because new software is added or software is fixed. Both of these changes affect testing and test procedures must take these changes into account.

Handling discrepancies

Discrepancies written as IRs are tracked in an automated system, and as software to correct the discrepancy is turned over to test, the status of the discrepancy should be automatically changed to "ready-to-test." This requires that the IR tracking system have an interface to the software build and delivery system. The tracking system must also provide a work list for each tester that shows all items the tester is responsible for. Testers retest discrepancies within a specified interval (e.g., one working day) and determine if the discrepancy is passed (IR is closed), failed (IR is kept open, responsible developer must fix it), or incomplete (IR is open and tester has additional testing before completing it). The incomplete status arises when the fix to correct the discrepancy could affect other functions. The tester verifies that the reported manifestation of the discrepancy is corrected but requires additional time to run testcases needed to verify that the fix has not adversely affected other functions.

Change of content

On a big system, several activities progress concurrently with the testing of the release that affect the release content. The activities that can affect the release content are:

- *Propagations*. If a change is made to software in a prior release, the change is propagated into the current release so that the new software does not lose this previously available capability.
- *Late delivery*. If software is delivered into the release after testing has begun.

- *Functional change.* Other testing, such as customer feature reviews, might identify a necessary functional change that causes software changes during the test period.
- *Development fixes.* As a result of testers identifying a discrepancy, developers change more than just the software necessary to fix the original discrepancy. While investigating the cause of the discrepancy, developers often find other discrepancies that must also be fixed.

Often, the content is changed after test planning is complete and the tester has determined what to test. A change in release content can impact testing greatly, making retesting necessary. The usual course of events is for testers to build confidence incrementally by running progressively more complex testcases. When a large change is introduced into the software, the procedures must allow for previously run testcases to be re-executed.

Reassessing test priorities

Periodically, the test manager must review testing efforts to ensure that testing will meet the scheduled completion date. If the software system is small enough, the status could be assessed by asking the testers how things are going. For large systems, the manager needs more specific data. The test procedures must ensure that the status information is available for each of the four types of testing—old functionality, corrections, modifications, and new functionality. The status of content change is also a valuable metric.

Old functionality status

The key item tracked to determine the testing progress for old functionality is the status of regression testcases. There are two factors that must be considered. First, the status of testcase maintenance, which should reflect the percentage of testcases that have been updated to work with the current release of the software. The testers should generate, as part of the release test plan, an estimated number of testcases requiring maintenance. The report showing updated testcases can be manually generated or can be an automated report that interrogates the testcase configuration control system.

The second factor is the status of testcase execution. As a testcase is run, its execution status should be automatically changed from "not run yet" to either "run successfully" (if no errors were encountered) or "run with errors" (if errors were encountered).

Correction status

An automated report should show the estimated number of IRs, the number found, and the number still open. A running history of these numbers during the release test period is also important to evaluate trends and to assure that progress is being made.

Modification status

An automated report should show the total number of SMR items and the number of closed SMR items. The status of modifications tends to be: few closed in the beginning, some closed in the middle, and most closed at the end. See FIG. 6-5.

New feature status

A report should be generated that shows what percentage of the test plan has been completed. This is a report about test status and should not be confused with a report showing what percentage of the features are working. Because the percentage of the features working depends on the implementation, it cannot be estimated by testers who are using a functional or black-box approach to testing.

Content change status

If the calculation is constrained to releases of the same system and to releases containing similar functional changes, it is possible to estimate the expected lines of new or changed source code by using history. An automated report can determine the actual number of new or changed lines of source code delivered from the software configuration control system and compare the actual number to the expected number. This quantified metric helps to assess whether the system is complete. If the actual number is lower than the expected number, possibly some expected software has not been delivered. If the actual number is higher than the expected number, possibly unexpected software has been delivered. Either of these possibilities mean that the system's content is not as expected.

Critical path

The release test plan should have identified the test's critical path, such as that item or collection of items that is estimated to take the longest time to complete. These items could have been assigned to one or more testers. If, for example, one person has the concurrent responsibility of testing a new feature, many SMRs, and old functionality that requires updating testcases, and this is the critical path, it might be hard to automate a way to measure this activity. Status is reported for each of the individual activities but not the combined activities. However, before making any predictions about the impact on the completion date, the critical path should be considered by the test manager.

After reviewing all the reports, the test manager can determine what type of testing needs emphasis, perhaps reallocating resources to address any testing that is behind schedule. These reports can alert test managers of the possibility of testing not being able to meet the project completion date.

Database procedures

This section addresses procedures for refreshing, converting, and updating databases. The contents of the databases and the necessity for control is also discussed.

Database content

The database must contain the data that is used for testing. It must contain the right type of data and the right amount of data. The requirements that the data must satisfy are:

1. *The data must support the testcases.* If a testcase uses data and does not restore the data before completing, then enough data of the type needed must be available to allow the testcase to run several (10) times. For example, if a testcase that checked a simple order that took something (a widget) out of stock and sent it to a customer worked, the testcase would require that several widgets be in the database.

2. *The data must support new feature testing.* Often, a new feature requires a new type of data or the addition of a new field.

3. *The data must be realistic data and special test data.* When testers are developing testcases by executing them manually, they use data that requires as few key strokes as possible. For instance, if a customer called *A* is entered as a customer in the customer database, then *A*, which is very easy to enter, can be used for entering most orders. However, testers must also use customer names such as "The Amalgamated American Distribution Network of South Plainfield" simply because some process downstream might handle a name that long incorrectly. Another reason for using realistic data is that the test database is often used to develop end-user documentation and end-users want to see realistic data.

4. *The database must contain old data.* Often, as a system matures, data that was once valid becomes unusual, and developers, while changing the software, overlook these unusual values. For instance, when IAM was first developed, all the orders were entered by the user and the customer order number was never known. Now, almost all orders, but not every one, are entered via a computer-to-computer interface and always have the customer's order number. When the order database is accessed, the developer cannot assume the existence of a customer's order number and must handle the case where it is absent. Data with the customer order number missing must be available to verify that the software continues to handle this case correctly.

5. *The database must have enough space to allow temporary test data.* When testing a new feature, a tester creates data that might be needed. When the tester develops testcases, the data requirements are determined and the required data is then entered permanently.

Database control

It is imperative that databases and testcases be in sync. False testcase error reports due to a mismatch could cause erroneous software quality evaluations that can potentially delay a project. For example, at the start of any testing phase, a sanity check is performed. If the mismatch causes all sanity checks to fail, then that phase of testing is delayed. Other database activities that require the database to be controlled are database conversions and new features.

When the structure, or global data content, of a database is changed from one release of the system to the next, the databases must be converted and the conversion tested. Otherwise, the scenario of convert, fix problems, convert, fix new problem, etc., becomes an endless process. The number of iterations and the timing must be controlled so other testers can use the databases.

Testing a new feature can often require new data that can only be entered using the new feature's software. The scenario is: enter data, test that the data is correct, correct problems, restore database to initial state, enter data . . .. Another endless loop. Unless the cycle is controlled, other testers never have a stable database. Databases can be controlled by using separate databases or by following a strategy for updating and refreshing the databases.

Using separate databases

Many of the problems described in the previous section can be eliminated if each tester has an independent database. The data content can then be changed without affecting other testers. Another way of providing separation is for a tester to use a separate portion of the database. This eliminates some interactions, but when a database is updated, however, it affects all the testers. Providing many databases would require an enormous amount of space unless the databases were downsized and contained only a sample of user data and the system must be tested with full, user-sized databases. While the use of separate databases does provide each tester with a stable database, control is still needed for user-sized databases.

Update strategy

Updating a database is a process of adding, changing, or deleting data that will be captured and permanently preserved. There are three individual actions:

1. Update—change the data content.
2. Backup—save the new content in an available but safe place.
3. Restore/refresh—return databases to their condition at the end of the last update.

Updating must be done both before the test period for new releases and during the test period. Updating is done before the test period for the release to enter the new data needed for testing the release. This supports database conversions and new features. Also, data can be entered for testcases that have been developed since the last release. The procedure is to refresh the database, then update during a short period, a window, then back up the database. If databases are to be converted, these activities must take place before the test period begins. During the test period, updates must be very short because updates affect all other testing activities. The window should be a half day at most.

Refresh strategy

Databases must be refreshed at a minimum: before the test period, at the end of the start-up phase, and before the final testing phase. Backups can be taken during the test period to preserve a hard-to-achieve database condition that can then be restored on request. If an extraordinary high number of testcases are executed, then a refresh might be needed to provide needed data.

Stress and performance testing

The environment needed for stress testing and performance testing is quite different than the test environment previously discussed. The hardware used must be as close to the user's hardware as possible. The system configuration should be the user's configuration. The databases must be sized and spread as the user's would be. For instance, a user might track a million orders that occupy three physical disks while testers use only one disk.

The object of stress testing is to verify that the system can continue to operate even though the expected input rate is exceeded. The rate of input is increased gradually up to the expected maximum and then beyond. Stress testing can discover a critical class of problems. If the system crashes because it cannot handle overload, which the user cannot control (i.e., the people who put orders in are at several locations and are not aware of any problem), the user considers this unacceptable.

The objective of performance testing is to verify that the system can furnish the expected level of service defined in the requirements. Performance testing is conducted with monitors (functions built into the system that record relevant performance data for future analysis) in operation so that any module that is using excessive resources, such as memory or time, can be identified and changed if necessary. The testcases used should emulate normal operations rather than unusual conditions.

Summary

Testing large software systems requires: determining what type of testing to emphasize, implementing and updating testcases, executing testcases and analyzing results, handling problems and software changes, and reassessing priorities. Procedures were described that assist in performing and monitoring these activities.

8

Test documentation

Test documentation is a necessary and useful tool for managing and maintaining big-system testing. The documents produced by testers answer the questions:

- What to test?
- How to test?
- What were the results?
- How can testing be improved?

This chapter provides testers with high-level guidelines for creating the necessary set of documents. If these documents are written using the suggested guidelines found in this chapter, their usefulness will extend beyond the current release, benefiting testers of future releases from the work expended and the experience gained testing the current release.

Test plans

Planning for testing is an activity that occurs throughout the software development life cycle. Plans for testing must be part of initial documents such as project plans or quality assurance plans. The following sections address several levels of test planning required for testing such as: the project, the release, and the feature.

GUIDELINE: The sooner testers are involved, the more likely that the final product will achieve its quality objective.

Project test plan

Test planning must begin at the same time project planning starts. Unless decisions about testing are made at the beginning, the necessary resources might not be available when testing starts. The project test plan must address the following areas: coverage objective, stopping criteria, resources, time, and staff.

Coverage objectives A specific coverage objective must be determined before other test planning can begin. What to test and how well to test are necessary inputs to determining how many people are needed or how much time is allocated. The coverage objective must address how completely to cover—the structure, the input domain, the output domain, the requirements, and the functions. Objectives in these areas are different during the software testing phases. Therefore, objectives must be specified for each phase, i.e., unit, multi-unit, product, inter-product, and evaluation. Coverage objectives depend on the project.

Software used in life-threatening situations or business-critical situations must be tested more completely. Still, ordinary business applications might require more coverage if the reputation of the company would suffer if the software did not perform well. Coverage objectives can be seen two ways: "The system will be tested this well," (positive) and "The system will only be tested this well," (negative). When developing coverage objectives, the cost to achieve a better coverage objective must be compared to the cost of correcting problems that would have been found with a better objective.

Stopping criteria The criteria for stopping testing must be established as part of the project test plan. The stopping criteria can be stated as "Less than X discrepancies have been found in the past Y days and less than Z of the open discrepancies affect user operations." These numbers, X, Y, and Z, must appear in the project test plan to prevent test managers from being pressured to stop testing so that the release can ship on schedule. Another factor that must be considered is software churn, or the number of modules that have changed in the past W days. If this is high, e.g., 2 percent, then the system has not have been stable long enough for testing to determine a good measure of the quality. [BEIZ87 DALA90]

Time The schedule for developing the system must include time for testing. Because each release of the system is different, the time required for testing varies. However, the project plan must still aim to set a realistic time span based on a generic release. A detailed time schedule is part of the release test plan, which is described later in the chapter. If the time allocated is too short, then the pressure to ship on schedule might cut testing short. One way to avoid this is to evaluate the progress of testing periodically, say every two to four weeks. If the predicated discrepancy identification rate or the predicted testcase completed successful rate are not as expected, then consider rescheduling the ship date.

Resources The project plan must address procurement of the following:

- Terminals or workstations.
- Computers for target environment(s).

- Time on off-premise computers or facilities.
- Test tools.
- Networks for local testing facilities.
- Networks for off-premise facilities.

Some of these resource needs can be addressed by using or sharing existing resources. Others might require development (a new test driver), or purchase (new workstations for additional test staff). Note that there is an interdependency between the resources needed and the staff.

Staff A plan must be developed to get testers. Experienced testers can adapt to a new system quickly while inexperienced people must be trained before becoming productive testers. The quality and flexibility of the available staff can sway other testing decisions. For instance, if testers have never developed a test tool, then it would be less risky to purchase a commercially available tool than to develop one. If the decision is made to implement requirements-based testing, then the test staff must be available early in the development life cycle. The number of testers required relates directly to the test coverage objective. The higher the coverage required, the more people necessary. If a decision is made not to test or to minimally test some portion of the system, fewer testers will be required.

GUIDELINE: If no project-level test planning is done, then testing is constrained to using existing tools and the methodology and strategy currently in use.

Release test plan

Each release of a system is different and requires planning. The content of the release dictates where testing must focus its resources, but other factors must also be considered:

Release content The release might contain new features and maintenance to old features. Each new feature requires a tester and a feature test plan (see Feature test plan on pg. 88). The tester might be totally inexperienced, inexperienced, or experienced. A testers level of experience determines how long a learning curve is necessary. The release test plan must allocate testers as needed to cover the features in the release. It also determines how much time will be required to test the release to achieve the coverage objective documented in the project test plan.

The release test plan can redefine coverage objectives, staffing requirements, resource requirements, etc. The specific content and outside commitments—e.g., the release was promised before the date it could be tested with current staff and resources—could cause changes to previously defined quantities in the project test plan.

Personnel growth Testers are more valuable if they are cross-trained so they can test many subsystems. The assignment of a new feature to a tester or merely switching the responsibilities of two testers can improve a testers view of the system. The release test planner should consider this option when assigning testers to features.

Additional responsibilities The reality of testing is that the job keeps expanding. An area that was ignored as not requiring testing now suddenly requires testing. The change can be caused by personnel changes in other areas. For example, a developer who tested a specific area in a prior release has since been reassigned or left the company and now the area must be tested by someone in the testing organization.

Risk The release test plan must assess and document the risks associated with the release. One technique is to determine the critical path, i.e., the feature that will require the longest time to test. If this feature is not available on schedule, testing should be extended. By periodically changing testing responsibilities, backup is available and the risks associated with a tester getting sick and delaying a feature or perhaps the release is reduced.

GUIDELINE: If no release planning is done, then how will new functions be tested?

Feature test plan

The feature test plan specifies the testing approach for a single feature in the release and is more detailed than the release test plan. A detailed feature test plan cannot be part of the release test plan because tester assignments are part of the release test plan. The feature test plan addresses:

- What to test.
- How to divide it up if it is a large feature.
- Who does the testing.
- Who are development and requirement points of contact.
- What tools and procedures will be used.
- What is the impact on existing testcases.
- What is the feature schedule.

All the physical pieces of the feature must be listed: the transactions, the screens, the databases, the batch runs, etc. for each of the various types of testing (e.g., sanity, screen, function, flow), the feature test plan defines the coverage objectives.

1. *Who does the testing?* If several people will be testing a feature, their specific responsibility must be defined in terms of the physical elements they will test and the levels at which they will test.
2. *What tools and procedures will be used?* A feature might require a new tool or testing procedure. The release plan should have already ad-

dressed acquiring the tool and the feature test plan must specify exactly how much the tool will be used. It is important to ensure that a new tool fits into the current environment.

3. *What is the impact on existing testcases?* A new feature can change the output of existing testcases. A big system could have more than 1,000 testcases in the regression test package and changing a significant portion (10 percent) is a task that requires planning.

4. *What is the schedule?* The schedule for achieving important objectives, such as the coverage objectives for the types of testing, must be developed. This key element of the feature test plan provides project management with the information to determine when a feature should be considered late because the objectives are not being met.

GUIDELINE: If planning is not done for a feature, the feature's effect on the release is an unknown factor that can cause testing to be late or incomplete.

The test environment

The tools and the procedures used to test the system must be documented. This enables new testers to become productive faster. It is difficult to commit testing resources to writing a test environment document when they are under the pressure of testing release after release and often concurrently testing multiple releases. But without documentation, new testers, or testers with new responsibilities, have a longer learning curve. They must learn from an experienced tester, and teaching takes time. The company benefits from having the test environment documented because it is easier to duplicate for a new project. The test environment documentation, especially the documentation of tools, must be a cookbook. A new tester should be able to read about a tool and use it effectively immediately.

GUIDELINE: Undocumented test tools and procedures are like undocumented software—unusable.

Test specifications

Each testcase should have a specification that would enable another tester to execute the testcase and evaluate the results. Testcases are used for many releases because their life span is the same as the application software. The cost of developing test specifications must be divided by the life span of the testcase. This serves as insurance for a tester being absent. It also allows for reuse of testcases, with slight modification, for other purposes. Testcase specifications should be within the testcase. However, for complex testcases, a separate, longer specification might be required. A new tester would rather develop a new testcase than try to figure out what an undocumented testcase does. To eliminate redundancy, test specifications should always be written.

GUIDELINE: Undocumented testcases cannot be used easily by other testers and might require a specific tester for evaluation. This is intolerable.

Test status reports

Test status reports provide management with a readout of the current quality of the release. Additionally, the test status reports must provide a measure of the progress of the testing effort. Chapter 11 discusses various ways of reporting test status in detail. Because producing these reports is an on-going effort, a report production tool must be part of the test environment. Most of the status information should be available from the automated test tools and the testers should not spend time collecting this information. Obviously, some status information cannot be automated because it requires tester input. For instance, the tester's view of whether testing is on schedule cannot be a report of testcases run successfully versus expected testcases. Because a single discrepancy could be blocking many testcases, the testers must be consulted.

GUIDELINE: Not producing status reports automatically and periodically will require that testers stop testing and produce a status report whenever one is needed.

Problem documentation

IRs are opened throughout testing to document discrepancies, most of which are closed before the end of the testing period. Some, because of various circumstances, are not closed before the end of the testing period and are deferred to a later release. These deferred IRs are low-priority problems that will not affect the user's ability to use the system effectively. They will be corrected when the code-in-error is changed for a higher priority problem or for a new feature. IRs are also deferred when the cost associated with changing the code to correct the discrepancy, and possibly introducing other discrepancies, outweighs the cost associated with shipping the system with a known discrepancy that can be corrected in a later release. This situation exists towards the end of the testing period when changing code might invalidate all previous testing.

All deferred IRs must be documented. Because IRs are release-oriented, they expire when the release is shipped and they can no longer serve as problem documentation. The customer is informed of the discrepancies by including a description of each deferred IR in the release package. The deferred IRs are then entered as new SMRs to ensure that the discrepancy is scheduled for a future release.

GUIDELINE: An undocumented problem will reoccur and time will be wasted reanalyzing it.

Test wrap-up

After testing for a release is complete, the testers must produce a test wrap-up. The purpose of this document is to report what went right and what went wrong. The wrap-up serves as a negative feedback loop to improve the process. The document must include: process problems, causes of schedule slips, causes of incomplete testing, and suggested improvements.

Summary

This chapter provides the "whys" behind the documents for: test plans, test specifications, test status reports, problem documentation, and test wrapup.

Test documentation represents a significant investment of tester time. By increasing the life span of documents to more than the current release, costs are reduced. The format of the documents is not as important as the purpose or the information conveyed. These documents must stand the test of time and be useful as the system changes. When viewed from this perspective, test documentation plays a key role in the testing effort.

9

Test metrics

Test metrics measure the quantity and quality of software, the progress of testing, and the effectiveness of the testing. Test metrics must answer "What is the quality of the software?" and "How well is the testing going?" This chapter concentrates on the derivation of test metrics. Attributes of a useful software test metric are: starts low, ends high (e.g., 100 percent), and is repeatable and comparable.

Software quantity

The amount of new or changed software in a release of a system is an important factor in determining how much testing time will be required and in predicting how many discrepancies will be identified. The change in the amount of software during the testing period is also a critical factor. Two metrics that measure the quantity of software are lines of source code and the number of modification requests (SMRs).

Lines of source code

The number of new or changed lines of source code is determined from the stored, source configuration control libraries. The algorithm used to count lines should be a company-wide standard that operates on any source language [PATT82]. The counter must eliminate comments. The expected lines of code is estimated before the release. As the test period progresses, the difference between the actual and the estimated lines of code is used as an indicator of the completeness of the functional capabilities of the system. The history of the number of lines of code is also important. If a significant change occurs (about 10,000 lines), the testers know that some new function is now available. If the actual lines of code exceeds the estimated lines of code, it indicates that the testing task has expanded beyond expectations and might require additional resources.

Number of SMRs

The number of SMRs in a release is also a quantity metric that should be measured. Because, on average, a tester can test one SMR a day, a rough estimate of the days required to test a release is the number of SMRs divided by the number of testers. (Note that the estimate of one SMR per day is based on the history of one project over many years.) The number of SMRs in the release changes during the testing period, which impacts testing. The number of SMRs increases by adding more propagation SMRs or by adding more maintenance SMRs. Deletions occur when the software is not available. The impact of these changes can be calculated using the following formulas: For each SMR deleted, calculate:

$$p_i = \frac{DD}{TD}$$

where: DD is the number of test days completed.
TD is the total number of test days.

For each SMR added:

$$a_j = \frac{TD}{TD - DD}$$

where TD − DD is number of days of testing left.

The Activity Impact (AI) is calculated as:

$$AI = \frac{\sum_{i=1}^{i=N} p_i + \sum_{j=1}^{j=M} a_j}{S} \times 100$$

where S is the number of SMRs at DD = 0.

When the activity impact is greater than 25, the length of the test period should be reevaluated.

Software quality

Because the purpose of testing is to measure software quality, the result of testing must be a metric that measures software quality. The quality of something is its basic nature in terms of the degree of excellence that it possesses in the eyes of the customer. The excellence of a product is established by performance. For instance, an automobile would be considered excellent if it performed trouble-free and did not require anything except ordinary maintenance. Even though the automobile might be test-driven, the excellence must be established by the owners. Simply verifying that each part meets specifications and that the parts were joined properly so that the final automobile meets designer specifications does not measure the excellence of the automobile.

The measurement is a necessary but not sufficient condition. The measure of quality that establishes a reputation is the quality perceived by the user. It doesn't matter if the automobile adheres to its specifications if the driver's feet can't reach the gas pedal. This example shows that there are two aspects to quality, one that can be measured before the product goes to the user; the other the user's perception of the product's quality after they use it.

One method used to control the quality of a product before it goes to the user is to inspect samples to ensure that they meet the desired standards. When a manufactured product such as a bolt fails to meet its specifications, it must be discarded. Perhaps the entire batch of bolts must be discarded. Similarly, software testers inspect sample executions of software to ensure the software meets standards and requirements. If the software fails to meet a standard or requirement, a discrepancy is reported. Discrepancies are also reported when, in the tester's judgment, the software does not perform in accordance with the customer's needs. Unlike the bolt that was discarded, when software does not meet its specifications, it is repaired.

This process of reporting and fixing discrepancies improves the quality of the software. Testing must be committed to finding any problem that will impact the user. Testers report the number of discrepancies found along with the percentage of expected discrepancies that have been found. To determine the percentage of discrepancies found, testers must have previously determined an expected number of discrepancies. This number varies by project and depends on the release content as specified by SMRs. These SMRs are classified for a release as:

Enhancement	change to the system
Maintenance	fix to the system
Propagation	previously tested SMR coming into the release from prior releases

Historically, the expected number of discrepancies for an on-line SMR versus a batch SMR has been different. The following table has been developed over many years for one large project. It can be used to obtain an initial estimate of the number of discrepancies that will be identified for a release.

on-line enhancement	SMR	–	1.985 discrepancies
on-line maintenance	SMR	–	1.125
on-line propagation	SMR	–	1.50
batch enhancement	SMR	–	.65
batch maintenance	SMR	–	.265
batch propagation	SMR	–	.08

This calculation is significant for releases with more than 200 SMRs. If the release is smaller, these numbers are used but with less faith. Historically,

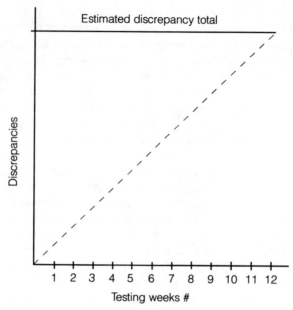

Fig. 9-1. Base-line for discrepancy discovery rate.

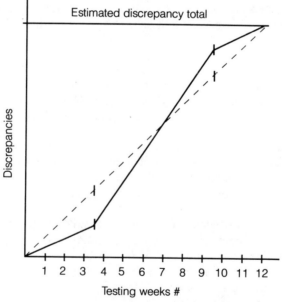

Fig. 9-2. Predicted discrepancy discovery rate.

testers on the project have found three discrepancies per thousand lines of new or changed source code (KLNCSC). The estimated discrepancies, calculated using the figures in the preceding table, can be divided by three as a check for the expected lines of code. The two estimates, discrepancies and KLNCSC, provide the basis for measuring how well the testing is progressing.

Figures 9-1 and 9-2 show how to build a predicted discrepancy discovery rate curve. The curve is divided into three parts: start-up, high discrepancy identification, and final. To begin, enter the estimated final discrepancy total and the estimated time period for test. Figure 9-1 shows these points with a dotted straight line, serving as a baseline, drawn between 0 discrepancies at time 0 and the total expected discrepancies at the expected end date (12 weeks, for example).

At week 3.5, mark the point half way between 0 discrepancies and the baseline. At week 10, mark the point half way between the baseline and the estimated final discrepancy total. These two points represent estimated points of inflection (knees) on the discrepancy discovery rate curve. This curve can now be approximated with the straight lines shown in bold (FIG. 9-2).

This prediction of how software testing will progress is based on historical data for a big system; data that has been recorded for many years and covers many releases. Developers are convinced of the credibility of this prediction because it has been within 10 percent of the actual values for many releases and similar curves have been plotted for other big systems. As one would expect, these predictions are revised during the test period if circumstances warrant. Factors that merit recalculating the prediction values are:

A significant number of SMRs are added to or deleted from the release   There are two major reasons for more SMRs: the number of propagation SMRs increases faster than expected, and the customer requests that additional maintenance SMRs be included. SMRs are deleted when the software is not ready.

Testing is not progressing as rapidly as predicted   The predicted progress of testing can be slower because too few resources were initially allocated. This under allocation of resources should be shown by metrics that measure test activities. But testing progress can also be slowed by discrepancies that are still open. First, individual discrepancies can further block testing. These discrepancies are categorized as critical and are normally fixed within two working days. Sometimes owing to other commitments, a critical discrepancy might not be fixed for several days and testing progress is slowed. Second, if a large number of discrepancies are open, testing can also be blocked. While no single discrepancy has a major impact, the accumulation of them has a large impact. Experience shows that if more than 50 discrepancies are open, test progress is hindered and when more than 100 discrepancies are open, testing usually stops! Progress, in terms of discrepancy identification, does not resume until retesting fixed discrepancies is completed.

The testing period is extended Late delivery of software due to development slippage or to user-suggested changes arising from the feature review can cause the testing period to be extended. When the testing period is increased, the predicted number of discrepancies identified is increased, not only because of the longer test period but also because the release size has increased. The rate of discrepancy identification might decrease but the total number of discrepancies will be greater. When a test period is extended, testers are able to increase the variety of input used to test new software as well as increase the coverage in lower-priority areas.

The reliability of any software-quality metric derived from testing is suspect if test coverage is not included. An absurd example would be running a single testcase, and if it worked correctly, declaring that the system quality is 100 percent. Let's make the example more realistic and increase the number of testcases to 500. If all the testcases work as expected, then software quality, derived for testcases, remains at 100 percent. But what if the testcases only covered 10 percent of the system's functionality? What then is the system's quality? Don't be trapped into simply multiplying functionality covered by the quality determined from testcase executions. Suppose the 10 percent of the system's functionality covered represents 80 percent of the system's usage. Then a reasonable metric should show that software quality is 80 percent. Therefore, the software-quality metric must be a product of testcase execution and coverage based on system usage.

Test progress

The progress of testing is determined by recording the status of:

- *Feature testing*. Feature testing is measured in terms of the percentage of the feature test plan complete.
- *Testcase maintenance*. Testcase maintenance is measured in terms of the percentage of the testcases that have been upgraded and run successfully versus the number of testcases that needed upgrading because the software changed.
- *Testcase execution*. Testcase execution is measured in terms of the percentage of testcases, from those selected for testing the release, that have run, run without errors, and run with errors.
- *SMR testing*. SMR testing is measured in terms of the percentage of the SMRs scheduled for the release that were tested and have: passed, passed preliminary test or failed, and those not yet tested.

The status for these items should be reported in two ways. First, in comparison to expected goals, and second, in comparison to previous releases with a similar content.

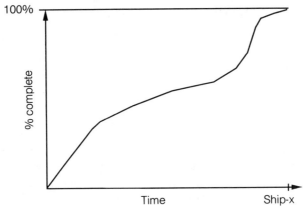

Fig. 9-3. Test progress.

Expected goals

Test progress, regardless of which of the four test progress metrics mentioned were used, has historically been nonlinear and often looks like the curve in FIG. 9-3.

The time, (ship minus x), is determined by subtracting from the ship date the time needed to package the release for shipping and the time needed for final testing of the release. The expected goals for the four test-progress metrics are interrelated because an individual tester could have responsibility for one or more parts of each metric. Therefore, the interrelationship must be taken into account when developing the expected goals. If the testing progress deviates significantly from the expected goal, it's an indication that the time period is too short or that not enough resources have been devoted to testing. If the time period is extended, then the expected goals must be recalculated beginning with the actual progress made to date.

Historical comparison

Comparing the progress of the current release with the progress of a past release with similar content, shows if the progress is within an envelop of expectation. Any significant deviation from this envelop must be addressed by extending the test time or supplying additional resources.

Testcase productivity

Automated testcases are deliverable items, and productivity measurements can be applied to developing automated testcases. Depending on the test driver, testcases can be code, even a significant amount of code, as much as

500,000 lines for a big system. The metric for measuring productivity is the standard:

$$\text{productivity} = \frac{\text{testcase lines of code produced}}{\text{number of personnel} \times \text{time}}$$

Productivity figures should be calculated at the end of the test period and then compared to previous releases. Significant differences between releases should be investigated by the test manager. A convenient metric to use is the number of testcases produced per week per tester. Low productivity (less than one testcase per week per tester) will reduce the coverage of the regression test package for succeeding releases. Too high a productivity (more than three testcases per week per tester) indicates that there is probably too little time being spent on planning for the next release. It could also indicate that the testcases being produced are clones of existing testcases with few changes and that there is too much redundancy.

Test effectiveness

Measuring test effectiveness requires a long-term perspective because test effectiveness is gauged by the SMRs from the field. SMRs are discrepancies that were not found by testing and, therefore, they indicate areas where testing can be improved. There is value in just using the numbers without looking any further into the SMRs. The rationale is that a "bad" system will have more SMRs than a "good" system. While this might seem to be more a measure of the software quality, it is also true that a "badly tested" system will have more SMRs than a "well-tested" system.

One metric often used to measure test effectiveness is field fault density, which is the number of field faults per thousand lines of new or changed source code. Conceptually, every SMR reported from the field per release is summed and divided by KLNCSC to determine the field fault metric.

Another interesting metric is field faults per system months. This is calculated by plotting field faults versus the number of production months, where production months is a summation of the months the software has been in production at each site. For a volatile system, a release might have many "sub-versions." A major release typically contains new functionality of a global nature while "sub-versions" would only contain maintenance modifications. For large systems, however, maintenance modifications can be 1/5 to 1/4 of a major release. To compute field faults, all the SMRs reported against any version of the release are summed and divided by the accumulated lines of code.

A truer measure of the testing effectiveness requires that SMRs be examined in detail. Some SMRs are reported by more than one customer. Duplicates can be eliminated when the emphasis is on testing effectiveness because there is only one discrepancy. Some of the field SMRs request added or new functionality. These SMRs could not have been found in testing based on the requirements and must be eliminated from the count. Some of the field SMRs are misunderstandings and are closed without any software change. These SMRs must also be eliminated from the count.

In the past, the ratio of where problems are found using data without adjustments is: [LIND88]

     80% before ship,
      4% in customer test, and
     16% in production use.

With refinement the ratio is:

     88% before ship,
      3% in customer test, and
      9% in production use.

The comparison between pre-ship and post-ship fault density is another metric to evaluate test effectiveness. The preceding figures show that the ratio of pre-ship fault density is approximately 6:1. If the ratio decreases, then testing is less effective. Again, SMRs that are duplicates, requests for enhancements, or misunderstandings, must be eliminated.

Summary

Metrics provide a realistic picture of software quality and testing progress. These metrics must be credible to serve as early predictors of resource shortfalls and warnings that the schedule might not be met.

10

Managing
the test process

Managing the test process is like juggling four balls. Each ball represents one testing component: regression, discrepancies, modifications, and new features. Management must define which of these components has the highest priority at any given time and change the priority as necessary. Preliminary planning ensures that resources for each component are available. As testing progresses, management reviews the status of each component to determine whether the initial resource allocations must be changed. Another way that managers can affect the rate of progress is to change the level of coverage. Test managers must focus the tester's efforts. Otherwise, testers will do the work that they find interesting. Another test management effort is showing the value of the job that testers are doing.

The software testing process

Managing the testing process is a balancing act. The software testing process consists of four separate testing components that the testers must address:

1. Regression
2. Retesting fixes
3. Maintenance changes
4. New features

Each of these testing components is equally important, and progress must be made in each concurrently and uniformly across the system. Imagine what chaos would result if, for instance, all the new features were completely tested and ready to ship and then regression testing started and the system no longer performed necessary basic operations because some of the new features were implemented incorrectly? This example can be repeated with any two of the testing components, all resulting with the same question—"Why wasn't this found earlier?"

Dividing the testing effort across each of the testing components will not avoid the question but will provide a reasonable answer, "It was planned and this was the time to find the problem. Otherwise, identifying other critical problems would have been delayed and the impact would have been greater." In order to do this, however, the manager must be constantly aware of each testing component's:

- Final objectives
- Current status
- Current problem areas
- Projected progress

These are all variables that change by component and by time. Even the final objectives can change. The next section covers how the priority can change during the testing period for each of the testing components.

Executing and maintaining regression testcases

Executing and maintaining regression testcases is an activity that occurs throughout testing. An initial estimate is made of how many of the regression testcases will require change and sufficient resources are allocated. Metrics are used to track how the effort is progressing. If the effort to run or to maintain the testcases is progressing slowly, it is critical for managers to reassign resources to increase progress. Managers must be wary of reassigning a testcase to a different tester, however, because it is just as difficult for a tester to become responsible for a new testcase as it is for a programmer to become responsible for a new program.

Any tester can assist another tester by performing subtasks, however. If the tester, called the primary tester, is not making fast enough progress in regression, then another tester, called the secondary tester, can submit and babysit testcases so that the primary tester only has to analyze failures. Another way a secondary tester can assist is by making global types of edits that the primary tester specifies to upgrade testcases. At the end of testing, the percentage of the regression test package that has run without error is a measure of the system's readiness to ship. If portions of the regression test package are not yet updated, the ability to estimate readiness suffers and the risk that the system will not perform correctly increases. If portions of the regression test package have not yet run, coverage is questionable. Therefore, the test manager must ensure that the regression test package is maintained and executed.

Retest IRs

When an IR is resolved with a change in the system, the system must be retested. All the other testing components generate IRs so that retesting IRs always occurs. One criterion for gauging test progress is the time that elapses before an IR is retested. It is realistic that IRs be retested within the next working day after they are fixed. One day turnaround may seem short, but by retesting quickly, the testing organization is overtly demonstrating a commitment to working together with the developers to improve the software. Fast retesting of IRs has a significant probability of rapidly identifying a new discrepancy because many years of experience has shown, an average of 30% of the IRs were fixed incorrectly or just allowed the tester to find another problem with the same testcase. Making and living up to a commitment to retest within one working day means that testers often have to change their plans. This is a constant source of irritation that testers must learn to cope with. Test managers must recognize that retesting has an effect on scheduled activities and may cause revisions of schedules periodically.

Maintenance changes

From a developer's perspective, most of the changes to the system are maintenance changes that affect only a limited portion of the system. They say, "It's only a one-line change." From a tester's perspective, the change must be tested in relation to everything else in the system that could be affected, which could include every screen and many batch procedures. Because the time needed to test maintenance changes varies, depending on how much is affected, the test manager must be sure progress is continually being made in this area.

If a tester cannot devote enough time to a testing maintenance change, he will most likely just verify that the problem is fixed and the function operates as expected. Testing other parts of the system affected by the change is neglected. This can lead to related parts of the system not operating properly. Not being careful enough can lead to not finding a second or third discrepancy that was masked by the first one.

Many maintenance changes are familiar to the tester. They are changes that were in previous releases or the problems were found by the tester. This can lead the tester to let some things slip past because he is not looking carefully at other areas that will be affected. Usually, a tester sees a change as something new that must be tested carefully, but when the tester is familiar with the original error, he will sometimes spend less time than necessary checking the fix. This is especially true in cases where the tester has responsibilities in other areas and time is short.

One way to ensure that enough time is allocated to testing maintenance changes is to include, in any time estimates, one day per maintenance change per tester. Often, the time to test these changes is lumped together with any regression testing and insufficient time is allocated.

New features

New features are the most visible part of a release because the added functionality represents the major selling point of the release. Because of the new feature's importance, experienced testers are assigned to test them. Feature test plans are then developed and reviewed, and developers are available to fix problems as they are found. Developer testing of new features concentrates on ensuring that they work as designed. Testers concentrate on new features because they learn a new part of the system. These two statements imply duplicated efforts in testing new features but each effort has a distinct value. The more developers test, the earlier that the tester can start executing productive testcases that do not find a discrepancy in the first step.

After the tester learns how the system operates, they identify or develop data, and then they develop testcases. Unless the tester begins these activities early, typically during the requirements phase of the feature, there will not be enough time to completely automate the new testcases. The tester will then execute testcases manually, reducing the probability of identifying discrepancies. While tools exist to automate testcases from manual executions, the resultant testcases are not necessarily automated. A "well-designed" testcase includes specific data and is repeatable. An automated version of a manual testcase could use any convenient data and is typically not repeatable.

A trap some testers fall into when testing new features is that in their desire to learn more about the system and to design and develop new testcases, they expend all their time testing the new feature. The test manager must ensure that the depth of coverage is not overly exhaustive and that the testers balance their testing efforts among the four testing components.

Setting priorities

The testing manager must reevaluate the priority associated with each component of testing during testing. If progress slows in one component, the manager must increase that component's priority. If progress is rapid in another component, the manager can decrease its priority. This is an on-going process that must be done periodically, weekly for example. The status of each test component can change rapidly and, if not monitored frequently, can cause a large slip in the system's schedule.

Several variables can be changed to speed up testing, but each has a risk associated with it as described later in the chapter. They are:

1. Depth of coverage.
2. Regression content.
3. IR turnaround.
4. SMR checking thoroughness.

Planning

The key to managing the testing process is planning. Without a plan, testers can only respond to the latest crisis. During testing, there will be many crises, and if time is spent on each, it will be impossible to make progress in the four testing components. There are two types of planning that are discussed in the following sections: long-range planning and short-range planning. These plans are test management plans and cover who will do the job and what are the possible contingency plans if priorities change.

Long-range planning

Long-range planning is the "nitty-gritty" details of the formal test plans discussed in chapter 8. The test manager is responsible for documenting the plans for testing a release in the Release Test Plan, which specifically identifies testing responsibilities. The test manager must consider how an assignment will affect the tester assigned. Some of the factors that must be considered are:

1. *New functions versus old functions.* From the manager's viewpoint, it is best to assign an experienced tester to test a new function because less learning is needed before they start producing usable testcases. Testing a new function can also provide a tester with new exposure and new recognition, thus helping to increase their standing. On the other hand, experienced testers might want to continue testing an old function. They are comfortable, they know all the people associated with the old function, and the tester might also want to complete the job. This is not a bad situation for the manager, because a tester responsible for an old test function should progress quickly and could be a valuable resource to shift to another function that is behind schedule if necessary. However, assigning an experienced tester to an old function is usually viewed as a person doing the same old job. The manager must determine the best time to move a tester. The manager should, when possible, assign a new tester to an old function where the prior tester can serve as a mentor. If a new tester must be assigned to a new function, the best situation is as a team member so other testers can serve as built-in mentors.

2. *Lead tester.* A team is sometimes needed to test a new function, and every team needs a leader. This position is called *lead tester.* The lead tester can be responsible for dividing the testing into smaller pieces, assigning them to individual testers, and overseeing and mentoring the testers. This position affords a tester the chance to learn management skills and interact with managers in other areas such as development. Because the lead tester position is slightly different, a tester can approach it with new enthusiasm and, thus, their performance is

improved. The test manager should be aware of the advantages and the possible drawbacks that increased exposure might have on an individual.

3. *New testcase development versus maintenance*. The test manager must assign testers to develop new testcases for new features. The development of new testcases is more challenging than maintaining old testcases and also has the advantage, to the tester, of producing a product. Maintaining existing testcases is vital, but it is also time-consuming and somewhat boring. A newer tester is usually more productive maintaining testcases but can overlook problems, therefore, a mentor is necessary. The test manager must be certain that testcases are being changed correctly and not just made to work.

4. *Resources for each test component*. When release planning begins, the manager must assign a tester to each component for each test area. Usually, areas are more numerous than testers so testers are given multiple responsibilities. When the work load on a tester is excessive, additional testers are required. If new people are not available, the manager must determine how to reduce test responsibility. Some possible actions are discussed later in the chapter.

Short-range planning

During testing, planning must continue but the focus shifts to "How do I get the release out?" In order to do this, the manager must be aware of some of the problems in the testing process.

Development problems   If development is behind schedule, software is delivered later than planned. Although the test manager must ensure that project management is aware of the tardiness of the software, he must also develop a plan that refocuses testing emphasis to bypass affected areas.

If the requirements are inadequate, then the code developed will also be inadequate and many discrepancies will be found. Other causes of large numbers of discrepancies are inadequate developer testing or just plain bad developers. Identifying an unexpectedly high number of discrepancies will slow testing progress and the test manager may have to adjust the schedule to compensate for "bad" code.

If developers cannot fix discrepancies in the prespecified time, testing slows down. Often, the next discrepancy cannot be found until the current one is fixed. The test manager must be aware of the turnaround time to fix problems and take every possible step to ensure that this time stays within the expected limit. If not, the manager must adjust the schedule.

Build problems   Software passes through a build process (i.e., BUSY) before being delivered to test. If the build process is delayed, then testing is slowed. The test manager must monitor the time required by the build process and take steps to correct the situation if it is too long. Software builders have many other products to build and the priority of any single product might have to be adjusted periodically.

Test problems   Problems in the testing process are usually seen in the test status discussed in the next section. The test manager must be aware of any area that is not meeting its predefined goal, determine the cause, and take the appropriate actions to get testing back on track. An easy-to-overlook problem is when two testers are testing the same item because of a misunderstanding. It is important that the manager talk to each tester periodically and make sure that they are not duplicating another testers tests.

Tracking the test process

A test manager must report to project management about the status of the test process. If the system is big, the task of keeping the status current is large enough to require automated tools (see chapter 9). One initial task of tracking the test process is to establish time-related goals. Goals such as the following must be set:

- Testcases run.
- Testcases updated.
- Testcases working.
- New testcases.
- Discrepancies detected.
- Coverage percent.

Another task is to establish some absolute goals such as:

- Coverage.
- IR turnaround.
- Software availability (if phased releases).
- Discrepancy delta, the difference between the number of discrepancies found and the number fixed.

The manager is responsible for determining goals, generating status reports, and determining whether any action is required based on actual performance compared to goals.

Management options

The test manager has some options when testing is not progressing as planned. He can speed up testing in several ways or can try to change the schedule. In either case, there is an associated risk that must be assessed before taking any action.

Resource shuffling

If work in one specific testing area is not progressing as planned, the people responsible for other areas can be shifted temporarily to provide a boost. The people chosen must be experienced enough with the area so that they do not need training. Another possible source of temporary help is developers or requirements personnel. They can ensure that the system operates as documented. However, it is extremely difficult for them to follow a predetermined testing sequence or to produce documented results that are usable enough to generate testcases. In an emergency, however, any help is better than no help. The risk associated with shuffling people is that the job they were assigned to before the move will be delayed.

Coverage goals

If the test manager reduces the test coverage goals in specific test areas, the progress towards the coverage goal is recalculated and the areas seem to be ahead of schedule. The reduction serves two purposes. First, the testing of the system as a whole is now closer to completion. Second, the testers responsible for the affected areas might be far enough ahead to help testers in other areas who are behind. When reducing coverage goals, the manager must assess the risk that even though the new goal is achieved, there could be problems that are not discovered. These problems would have been discovered if the old, higher, goal had been used. The cost of fixing the undiscovered problems is the measure of the increased risk.

Regression content

If an area has not changed and has not been impacted by change, the regression coverage goal could be reduced by reducing the number of testcases in the regression test package. The probability of finding errors is less in this area and the time saved might be worth the risk. Also, if an area is new, the developers probably have tested the area extensively and it might be possible to reduce the coverage goal to a lower level and rely on the developer's testing to have found most of the problems.

IR turnaround

The test process can also be speeded up by shortening the IR turnaround time. The risk associated with a shorter turnaround time is that less developer analysis will be performed and a higher probability that a fix will introduce additional discrepancies. Conversely, by increasing the IR turnaround interval, testers will not be able to identify as many additional IRs and the discovery curve might appear to flatten, e.g., "The system looks ready to ship," when it is not.

SMR thoroughness

The test manager can reduce the depth of testing SMRs. This can be done in a variety of ways. Any propagation SMR that has been released before could be tested indirectly by exercising the system and assuring that nothing unexpected happens. An SMR could be tested by only one testcase that exercises the system in exactly the way the discrepancy was discovered and assuring that it is fixed. This eliminates executing the many other testcases in different areas of the system that could be affected by the change. The risk associated with reducing the thoroughness of SMR testing is that the user will find another discrepancy related to the SMR being fixed and believe that testing was insufficient and that the product quality is inferior.

Schedule slip

If the risk of shipping the software without fixing the known discrepancies is too great, the ship date must be slipped. The test manager must determine how much of a change is required. This is not a linear function because other resources, e.g., requirements people, developers, and computers have new commitments and might not be available for the extended test period. A schedule slip is also not easy to sell to product managers who are concerned with meeting schedules. Customers do not like schedule slips either. They have already committed resources based on the previous schedule. The test manager must provide early warning that testing is not progressing as predicted and be able to quantify the difference in terms of problems identified between shipping on time or slipping the schedule.

Although a schedule slip might be suggested, the product manager might decide to ship regardless because the cost of missing the ship commitment might be greater than the cost of fixing the problems in the field. The test manager assesses the quality of the product and tells the product manager. The product manager makes the ship decision based on this and other pertinent information.

Summary

Managing the test process must include:

- Initial assignments to adequately cover each of the testing areas.
- Shifting resources to cover areas that are progressing at slower rates than expected.
- Reducing coverage to speed up progress.

Test managers must be prepared to report and explain the status to higher management.

11

Managing management

Like other managers, project managers want to know how testing is going. However, they become more nervous when their product is being tested. Product test is the last step in the development cycle before the product is shipped, a key milestone for the project, and testing reports the quality of the software. Project managers do not want to hear that the quality is unacceptable! They have been working hard to develop a high-quality product. Consequently, if testers do not quantify reports and do not appear to be trying to solve problems, then managers can view testing as an obstacle. Testers must report status. They must be honest, but they must also be concerned with the effects of negative reports. They must minimize the impact with an obviously cooperative attitude.

Who are they?

When a system is being product tested, suddenly a host of people become concerned. "How is testing going?" and "How good is the system?" is often heard. The people asking these questions are not only the developer and requirements people that testers are used to dealing with, but project managers, product managers, and people working on systems that interface with your system. They ask questions and expect answers, and if the answers are not what they expect, they take some unexpected actions. For instance, if they believe that the product is behind schedule and will impact their project, they might raise a jeopardy that is escalated to, and must be resolved at, high-management levels.

Why do they press for status?

Project managers want the product to ship so it can start generating profit. Their goal is to ship the product on time because their performance is judged by making that critical date. From their viewpoint, shipping on time is the only milestone to meet, even if no prior milestone was met i.e., testing must finish on time even though the software was not available on time. They are anxious and nervous about the status and they especially need to know if the schedule must be adjusted. The earlier this need for adjustment is recognized, the more credible the action is in the eyes of the customer. They do need to know, and they will exert pressure to receive the information they require.

They want quality reported— but it better be good news!

Project managers and other managers asking about the status of the system want to know about the quality. Their goal is to furnish a high-quality system, one that satisfies the user's needs in a timely manner, and they want to know if things are going well. From their viewpoint, they have done everything they could to ensure that the system was on time and would satisfy the user. Now they hold their breath and wait to see if the measurement, determined by the testing process, finds that the system will satisfy the user. If the quality of the system is not as expected, then all prior efforts might have been inadequate and they must now regroup and develop new plans. At this point, the last step of the life cycle, this news is not appreciated.

They want jeopardies and action plans

If system testing is not progressing as expected, project managers want to know, but they need specifics like what deliverables are in jeopardy. Because the system is big, they need a way of focusing on small areas they can manage. For instance, if the discovery rate of IRs is below expectations, testers must narrow the report down to pinpoint a specific component or feature. Otherwise, the area is too broad to focus on. Reporting a jeopardy is only the first step. A plan of action must also be developed. Managers can respond to plans by implementing them or changing them. Without a way to focus on a specific problem area, managers sometimes focus on the testers as the problem. A "kill the messenger" reaction.

Test managers must determine where extra effort would help and ask for it. Therefore, when you single out one area, you must be certain. All the predictions are based on large numbers so that local effects are washed out. By concentrating on one area, the washing out of local effects is diminished, and you must carefully consider the possibility that a specific area might behave differently from the system as a whole. You must also consider that singling out one area of the system could have unexpected results.

The impact of scrutiny

When the test manager focuses on one area, be prepared for a reaction. Managers, developers, and requirements personnel do not appreciate being singled out as what they perceive is the cause of the system being in trouble. In all fairness, one group usually is not the cause. Most often, it is a chain of events that culminates in the test phase. Still, singling out one area causes reactions you should be prepared to handle:

The specifications were no good or were late Developers do not want to bear the blame and will often point out that their input, the specifications, were either late or were not adequate. Testers must avoid the issue of "Who is to blame" and concentrate on "How do we improve the system?"

The software is okay in my environment The development environment should always have newer software than the test environment. Also, the development environment is not as tightly controlled as the test environment, and things often work in the development environment that do not work in the test environment. The developers are trying to avoid getting a discrepancy against their code. A tester could appeal for help to determine why the difference exists rather than insisting that the software has a fault.

The discrepancy was identified too late If a discrepancy is identified late in the test cycle and the priority is not high or a large amount of code must be changed to fix the problem, then the problem could be deferred to the next release. Sometimes, the probability of causing a new discrepancy that might not be discovered is high enough so that the system is better with known problems than unknown problems. In these cases, a tester should be sure that the priority is not high and that, if a work-around exists, it is included in the release package. If the tester believes the priority is high, then the tester must get support from requirements personnel or users to convince the developers that fixing the discrepancy is more economical than causing an extra maintenance release later or, worse, causing the customer to use cumbersome work-arounds.

Developers can hide fixes If developers want to keep the number of discrepancies lower, they fix more than one discrepancy at a time. They turn them over as a package to testers and then say they cannot duplicate the "hidden" discrepancies so they should be canceled. A tester is somewhat handcuffed trying to prove what happened and it probably is not worth the effort.

Developers can stop cooperating Testers and developers form a team. If developers feel threatened because their area is in trouble, they might react by pulling away from the team. A tester must make sure that the developer does not take things personally, but that they believe that, by teamwork, all can improve the software.

What will not sell

Management wants the testers to be objective. Even though testers have been doing the job for years and have developed an intuitive feel for what the status is, this intuition cannot be used as the basis for status. It will not sell because it cannot be quantified. But by showing historical comparisons, the message gets across. The last thing any tester should ever say is "I told you so." This is almost always destructive, even if it is true. All this says to management is that you did not do a good job of selling your point of view before. You can refer to predictions without ever saying, " I told you so." Again, gut feelings won't work, but one interesting way around this is to classify discrepancies into where they are expected to occur during the test interval. This would enable a tester to say, "We are finding too many phase I type discrepancies to be in phase III."

Summary

Testing must determine and report the quality of the software to management. The report can contain bad news, but it must always address how to improve the quality.

The status reports in chapter 9 are necessary but reports to management must contain additional information. Project managers want to know what the jeopardies are and they want action plans that focus on the area in jeopardy.

Reports that focus on an area can cause negative reactions, therefore, be wary. Criticism without an objective foundation will invariably backfire. Constructive criticism, including plans to improve quality, are viewed positively.

12

Automating the Testing Process

Testing a very big system can only be done using automation. The tools must be integrated and must be easy to use. The areas for which tools are needed are:

- Content control
- Test execution
- Test maintenance
- Test metrics
- Documentation

Appendix D provides detailed specifications for several of these tools. The tools discussed in this chapter might seem to overlap with tools used by developers or requirements writers. The cause for the apparent overlap is that testers need the information from the same sources.

System content

A big system has many parts and keeping track of all the changes, additions, and deletions cannot be done manually in a timely fashion. Tools must be developed and used that automatically track the system content from the beginning of testing to the end. Without knowing the content of the system, there can be no coverage measurements. Without knowing the changes, there is no measure of fault density. Without these measures, there is no way of determining the risk associated with shipping the system. Several tools are described in the following sections that are needed to determine and track system content.

The Initial Contents tool

The purpose of the Initial Contents (IC) tool is to define the contents of the release in terms of the SMRs that are included. The IC must have the capability to receive SMRs from many sources: users, requirements writers, developers, and testers. It must allow SMRs to be grouped into releases and allow SMRs to move from one release to another. If an SMR is needed in multiple systems, e.g., the code to be changed has been reused in different systems, then the IC must provide separate instances of the SMR for each system. If an SMR requires work by more than one development group, the tool must support multiple, responsible developers.

The IC must produce reports by test area and by development group. It must track all SMRs for one product, and to eliminate duplications, it should be able to track SMRs for all products that use the indicated software. However, the IC is limited to providing an initial view only and cannot be used to track the SMRs once testing is started.

Each SMR contains a description of the change and identifies all the parts of the system that will be affected by the change. The SMRs also contain the date entered, the scheduled release date, the responsible developer(s), and a size indicator, such as predicted lines of code. IC's output is reports and a file containing the SMRs included in the release. The file is used by the tool to track content changes as an initial starting point.

Reports by the development group and test area are used by the test manager to determine initial tester assignments. The predicted size of the release, the sum of all the SMRs, can be used to determine the length of test time and to predict the number of IRs that will be found. The IC must also produce reports that are date-oriented so they can be used to ensure that SMRs are addressed in a timely fashion. For example, how many SMRs are open more than 60 days. Keep in mind that IC is very important for keeping track of SMRs throughout the development and requirements portion of the software development cycle. From a tester's perspective, it only provides the initial content of the release that must be tested.

The Change Tracking tool

The purpose of the Change Tracking (CT) tool is to track all changes to the system during the test period. It tracks not only SMRs but also IRs. CT's input is the initial content file from IC. CT must be able to add SMRs, delete SMRs, and change SMRs. It must also take IRs as input.

The CT tool tracks SMRs and IRs by status, enabling the tester to change the status of any item manually and, in many cases, changes the status automatically to maintain consistency. For example, the initial status of an SMR is ST (in System Test) but if an IR is written against the SMR (a testcase failed), then the status is automatically changed to STF (System Test Failed). When the software is changed to correct the IR, the IR is retested. If it is correct, then the IR's status is changed to STP (System Test Passed) and the SMR's status automatically returns to ST. Changing the IR's status could have been manual or an interaction with a test execution tool.

The date of any status changes must be saved and provided as necessary. CT must provide information to metrics generation tools. The data contained in an SMR and an IR is similar; each contains a description of the problem, dates, and responsible tester information.

The reports that CT must generate include:

- All SMRs and IRs.
- A subset of SMRs or IRs selected by specific status.
- SMRs only for a selected development group or by tester and by status.
- IRs only by development group or by testers and/or by status.
- IR descriptions for IRs not yet fixed.

A final report should be available that provides IR descriptions of the IRs deferred to later releases. These reports are used to track the progress of testing and provide the test manager with a way to determine if an area is behind schedule.

The Build tool

The Build (BL) tool builds the system from source code after the developer delivers the code to BL. BL constructs all the libraries (data sets) needed to execute the system in the test environment and sends them to the test environment on request. This tool must provide a source control function. It must be able to associate related software and build those that are interdependent whenever any of the dependent pieces are delivered. BL must be able to provide a list of interrelated parts without executing any builds. BL must be flexible enough to accept any language's source code and assemble/compile/ manipulate it as required.

The BL tool should validate that each item being processed has an SMR, IR, or both associated with it. BL automatically changes the status of the SMRs and IRs to show that new software has been delivered and can be tested or retested. BL's input includes any type of source, even files that contain documentation.

The reports generated by BT include:

- Daily activity of what was delivered.
- Whether it was built successfully or not.
- The last build date of any part of the system.

Using the activity files, BL should produce a report of the expected SMRs that have not had anything delivered. The reports inform the test manager of the exact status of the system.

The System Definition tool

The purpose of the System Definition (SD) tool is to produce lists of all the other parts of the system that are not defined by the preceding tools from a single source. SD must be capable of maintaining lists that can change and associating the lists with releases of the system. The lists must be maintained by release. Some examples that illustrate SD's abilities are:

1. *A list of all the transactions in the system.* An IBM IMS system has a list of transactions in the system gen. SD must provide a method of reading gen input, selecting all the transactions, and creating a new list with changes (additions and deletions) noted.
2. *A list of all the commands in the system.* When a system is command-driven, there are extra commands that can be executed internally for maintenance purposes. The tool must be able to extract only the external commands and provide a list for a new release of the system with the changes from the previous list marked.
3. *A list of databases/data sets.* Databases are usually defined externally to the application. A system could define all the data sets internally by creating them when necessary. The tool must be able to extract a list of databases/data sets and provide the list with changes from the previous list marked.
4. *A list of the transaction usage profile.* The usage profile is a key element in determining the importance of testing for a given transaction. [MUSA87] The profile is provided by the user after the system is operational. An initial list for a new system must be developed by requirements writers. SD must accept a new profile as input to update a previous one.

SD does not interact with other tools but provides reference data. The data it uses are the lists it maintains. The output is a set of lists and changes to the lists organized by release. Testers use these lists to ensure that they have covered all areas of the system. In the case of the usage profile, testers can ensure that heavily used areas have been tested more than less-used areas. Testers can also use information in the profile to prioritize problems.

Testcase implementation and maintenance

Testcase development and maintenance are time-consuming activities. To increase tester productivity, it is essential to use automated tools to help develop and maintain testcases. Many of these tools are the same ones used by developers to implement and maintain software. Some, however, are unique to testing.

Configuration Control tool

The purpose of the Configuration Control (CC) tool is to maintain a central storage area for testcases, testcase specifications, and testcase reference data. It must be able to replace a member with an updated version and to furnish any version of a member to enable a tester to update the member. It must be able to add new members and delete old ones. CC must interact with the planning tool to validate that a testcase is scheduled to be added, changed, or deleted.

When a testcase is changed by a tester, CC notifies the planning tool of the latest level. Each member controlled by CC will have a header record that contains information to allow the member to be retrieved or listed by:

- Testcase ID
- Specification ID
- Test area
- Test type
- Release
- Responsible tester.

CC provides the latest, or any prior level of the source, and enables a tester to update any member without destroying the prior copy. It also provides lists of currently available testcase IDs or specifications IDs, with attributes that are useful for comparing with the planning tool output.

The Planning Tool

The purpose of the Planning Tool (PT) is to furnish a single, master source of testcase information and status. PT must provide the ability to maintain a list of testcases with status-related information. The tool must be able to produce a list of testcases by various selection criteria—release, test area, tester, test type, items tested, status, or combinations of these values. The PT tool must allow testcases to be added and obsoleted (the entry must remain but the status changes to obsolete). PT should be interactive to allow a tester to select a testcase or list of testcases and then change status and provide dates and update the master list.

The master list maintained by PT is used by the configuration control tool and by tools described later, i.e., the specification generator, the cross-reference tool, the executive, and by test status. Each testcase contains descriptive data about the testcase and status data. The descriptive data describes what the testcase will verify, e.g., the release, the test area, the transaction, the command, the screen, etc. The testcase status describes the current state of the testcase—working, being planned, being tested, etc. The master list of status allows the test manager to determine how testing is progressing with respect to new testcase development, testcase maintenance, testcase execution. It also provides each tester with a work list of the testcases they are responsible for.

The Specification Generator tool

The purpose of the Specification Generator (SG) tool is to increase tester productivity and to automate the creation and modification of the test specification. SG must be capable of creating and sending to CC, the configuration control tool, a new test specification. It should work interactively and supply prompts for the tester to enter the required information.

As a user-friendly additional feature, SG remembers the last response the tester entered and furnishes it automatically, so if there was no change necessary, the tester could just proceed to the next field. When all the data has been entered, the tool should send the latest specification to the configuration control tools and notify the planning tool that the specification has been completed.

SG must interact with the configuration tool to retrieve a previous copy of a specification and to store the updated/new copy. SG must also interact with the planning tool to update the status of the specification. The data contained in a test specification is described in chapter 5.

The output of SG is a specification that is used as documentation for a testcase. The need for testcase documentation when there are thousands of testcases and when test personnel change periodically is critical.

The Testcase Development tool

The Testcase Development tool provides testers with a single environment for developing and maintaining testcases. Testcase development and maintenance is the most labor-intensive task for testers. A single environment that provides all the necessary capabilities for testcase development will increase their productivity. The mechanics of putting a testcase together and changing it as needed must be as automated as possible. The environment is not a tool, but rather a collection and integration of some generally available tools, some specialized tools, and some procedures. The tools and procedures can be divided into three areas; general, development, and maintenance.

General tools   The items in this section are used when developing and maintaining testcases:

Separate areas   Each tester should have an independent work area so they can develop and change testcases without affecting the ability to run the regression package. This separate environment should contain their own database or portion of the database so that testers can establish test conditions to verify the testcases.

Editor   The editor must be part of the tester's environment. The tester should not have to move the testcase source from one place to another just to edit it.

Configuration control interface   The testcase source, specification, and if necessary, reference data, are stored in the configuration control tool. The tester must be able to retrieve a copy of the testcase and anything else at the latest level or at a previous level from this environment. When work on the testcase is finished, the tester must be able to archive the new or changed testcase without changing environments.

Status changes for planning tool   Because the planning tool maintains the current status of the testcases, the tester must be able to change the status to reflect their efforts.

Compiler   I use the term *compiler* as the name for the tool that prepares a testcase for execution. The compiler is not limited to transforming source into object code. A compiler could also be a testcase generator that transforms requirements to input data, or a message processor that transforms a message in human-readable format to a machine-oriented format with execution-specific data added. Another example of a compiler is the recorded session when using a record/replay type testcase creation tool.

Execution   A tester must be able to submit a testcase to be executed and get the results back. The execution would be by the tools discussed in the next section. An option must be provided when submitting a testcase of whether or not to change the testcase status. For example, during the development of a new testcase, it might be executed several times to verify the subsections and these executions should not be recorded.

Because the compiler for testcases that are scripts must provide a set of capabilities that include the set needed by compilers for other types of testcases, the script testcase compiler is discussed in further detail.

A script testcase could have many elements that are called reference data. To explain these elements, let's refer to part of a sample testcase shown below. A script is a sequence of actions that would be performed at a terminal. For this example, we'll use a testcase for an IBM IMS system using formatted screens.

```
GET (ORDIN)
.order = "ORD 1"
.date = today
Press (FIND)
if (MSG) ≠ ordmsg.7 error (1)
```

This section of the testcase does the following:

1. GET is a function that discards pending messages, if any, clears the screen and types /FOR ORDIN at position (0,0) and hits the ENTER key.
2. Type "ORD 1" into the field with label ORDER.
3. Type the current date, supplied by the function today, into the field with label DATE.
4. Press program function key 1 that, on the format ORDIN, provides a find function.

5. Perform a comparison and if the returned message, contained in the MSG field at the bottom of the screen is not equal to the seventh entry of the ordmsg directory, then report error 1.

This small example shows the elements of a script testcase.

- *Format Definition.* A file that contains the tag position equivalence. This permits the use of a notation like .order, instead of (2,17), the screen position row 2, column 17 where the data for tag ORDER, is to be entered.
- *Common definitions.* A file that contains a program function key equivalence and the definition of position for common tags like msg. This permits the use of notation like Press (FIND) which is more readable then Press (PFK 1), and the use of a variable like MSG.
- *Macros.* The macros in this file provide the capability of performing a sequence of actions like those described above for the macro GET.
- *Functions.* This file provides the capability to use notation like "today," a function that furnishes today's date. The compiler must provide the capability to transform the source into object code. It must be able to retrieve any of the elements from the configuration control tool if the element does not exist in the tester's environment. This allows a tester to obtain a new version of an element and test it before archiving the element for general use.

Output capture   An output capture tool must be available that provides the capability to save the output from a testcase execution. The tool must also prepare the captured output for comparison. For instance if the output contains the date of execution, each execution would contain a different date. The front end of the capture tool must be capable of eliminating the comparison for the date field or changing the date to a default value.

Database utilities   Testers need the capability to establish the database conditions necessary for testcase execution. The capability must be provided to refresh the database to an initial state or a previously saved state. The capability must be provided to alter the database, e.g., change the value of a field, adding a field, deleting a field, adding a record, or deleting a record. These capabilities enable the creation of test conditions that the system could not create during normal use. These records might be created accidentally or by faulty software. Examples are an invalid value or an illogical record.

Message formatter   Messages, either input or intersystem, tend to be compressed and unreadable. For instance a message could be:

```
*CTL{ORDIN;IAM;CUST;2681990}*ORD{..CUSTOMER = 17265;TYPE = N...
```

The formatter would convert this message to a more readable or beautified format:

```
*CTL{
      ORDIN
      IAM
      CUST
      2681998
      }
*ORD{
      CUSTOMER = 17265
      TYPE = N
```

The formatter should also be able to convert from the beautified format to the compressed format. The beautified format is also much more manageable.

Development   The items discussed below are used for developing test-cases.

1. *Prototypes*. If it is possible to produce a testcase prototype then it should be done. Each new testcase would start with the prototype as a skeleton and then more information is added to create a full testcase. The use of a prototype furnishes a single point of control so that a change to the prototype is propagated to all new testcases when the tester starts with the current prototype.

2. *Makefiles*. If a testcase contains elements like input files, output samples for comparison, etc., then some method must be provided to find and combine these elements both for compiling the testcase and for executing it. Makefiles provide this capability by processing the testcase source and combining the results with tester-supplied information to create a file. The makefile is used at compile time to retrieve the latest version of elements like format descriptions. Makefiles are also used at execution time to retrieve the latest version of a captured output for comparison.

Maintenance   The items described below are used to maintain test-cases, i.e., to update the testcase so that it works with a new release of the system.

1. *Global change*. A method must be provided to change all testcases in a specified manner. An example would be if testcase output is directed to logical terminals and the names of the terminals are changed. All the names must be changed and the testcases compiled again if necessary.

2. *Subset change*. A method must be provided to change a subset of the testcases in a specific manner. An example would be if the testcase

refs to some tester-specific item like directing information via a tester's login ID and a new tester is given the responsibility for the testcases.

3. *Record/replay changes.* If the testcase is a replay of a previously recorded session of a person using the system, then some changes in the system might render these testcases invalid. For instance, if the system uses formatted screens and the formats change, then the testcase might be attempting to put input into non-input areas. The format change must be analyzed and the replay file changed to correspond to the new format.

The Cross-reference tool

The purpose of the Cross-reference (XF) tool is to provide a cross-reference between the software components so that when one is changed all others that might be affected can be tested. XF must have the capability to maintain a cross-reference list by chaining together the identification of modules with the software that the modules affect or use. The chaining must be directional forward, backward, or both. The cross-reference tool must accept as input a tree structure from the build tool and be able to add to it by manual input. If a new tree structure is input to represent the latest release the tool must be able to merge its list with the new one. XF must receive input from the build tool as an initial list and also must accept notification about any software changes. XF's output is a list of software to test when a change occurs.

Coverage tool

The purpose of the Coverage tool (CO) is to provide a quantitative coverage metric. CO must maintain a list of software entities that can be tested independently, e.g., a subroutine called by a module cannot be tested independently from that module. It must also maintain a list of testcases and the software that each testcase covers. When a testcase executes without errors, then the tool must update the testcase entry and all the software that the testcase covers. CO must be able to determine if any software entity is not covered. CO will be notified by the test execution reporter when a testcase executes and the result must be kept in CO's database. Testcase lists are furnished by the planning tool. CO will report software not executed, and thus not covered, so that the test manager can ensure that it will be covered. CO will report what percentage of the software has had testcases run against it and if the testcase results were error-free or contained errors. Coverage is an important metric in determining the software quality.

The tool chest

The tool chest provides a centralized storage area for the testers to develop and store their test tools. The tool chest should contain an index of each tool available with a description of how the tool is used. The reason for a tool chest is to provide a place to store the tools the testers create and to eliminate the duplication that would occur if more than one tester developed the same tool. Several examples of tools that might be in a tool chest are tools that:

1. Print out all comments contained in a testcase that is a script.
2. Print a testcase in an indented "beautified" format.
3. Collect all parts of a testcase and prints them. This could be used to provide a hard copy record of a testcase.
4. List the dates and results of the last 10 executions of a testcase.

Testcase execution

The purpose of the Testcase Execution Tool (TET) is to execute testcases and to report results. The description of the tool is divided into three integrated components: the executive, the drivers, and the reporter [DONN87] (TET is also described in appendix D).

Executive

The executive controls and coordinates the execution of testcases by the test drivers. The executive serves as the tester interface. The executive must be able to direct a driver to execute a single testcase or a series of testcases. Testcases can be composed of sections that are executed by separate drivers. The executive must be able to direct a driver to execute a section, then direct another driver to execute the next section after the first section completes with the expected result. The executive communicates with the reporter to provide and retrieve testcase results. The executive communicates with the drivers to determine the status of testcases and the state of the drivers. The executive must be able to reestablish itself if it aborts.

The executive must communicate with all the other components of the execution tool. By providing a single point of control, the testers only need to learn one interface, the executive. This eliminates the learning curve associated with using a new testcase driver for every system or every new environment.

Driver

Drivers execute testcases and send the results to the executive, which sends the results to the reporter. A driver operates in one environment for one type of testcase. All drivers operate under the control of the executive. They are told to execute a testcase or a section of a testcase. They then retrieve the testcase from the configuration control tool or from the tester's work area (a new testcase) and execute it. When execution is finished, the driver notifies the executive of the results and the executive notifies the reporter. Drivers must also be able to communicate their current state to the executive and also inform the executive when they are idle and can execute the next testcase. Several drivers for different types of testcases are discussed below.

1. Capture/replay—replays a previous session.
2. Programmed script—executes testcases that are programmed scripts.
3. Message insert—executes testcases that are messages inserted on communication lines.
4. Batch—executes testcases that are batch jobs.
5. Printer—accepts output normally routed to printers and analyzes the output for expected results.

Reporter

The purpose of the reporter is to store the results of testcase execution in one place. The reporter furnishes testcase information when, and as, requested by the executive. The reporter is notified by the executive of the result of an execution and stores this information. The reporter maintains a history of testcase execution during a test period. When a tester wants information about testcase execution, they interact with the executive. The executive queries the reporter and presents the information to the tester. By storing all testcase execution information in a single place, the status can be quickly retrieved.

Metrics generator

Generating metrics manually is a time-consuming task. Because the status of testing is most effectively described in terms of metrics, they must be generated. The tools described below automate the generation process, which saves time and produce consistent metrics that can be compared from release to release to give a historical perspective when necessary.

Test status

The purpose of these tools is to automatically generate the reports needed to describe the test status. The tool must access the information stored in several places and produce the following reports:

Incident report discovery and closure From the data in the change control tool, develop a graphical display of the history of how many IRs were found, how many were closed, and a reference line of the expected discovery rate.

Software maintenance requests closure From the change control tool, develop a graphical display of the history of the number of SMRs that have been closed and a reference line of total SMRs in the release.

Testcase status From the planning tool, develop graphical displays that show the history of the number of testcases executed, the number of testcases executed without finding IRs, the number of testcases executed that found IRs, the number of testcases that have been updated to the new release and are now available, and the number of new testcases and a reference plot of testcases available. The number of testcases available must increase or decrease as testcases are updated, added, or deleted.

The tool, as previously discussed, accesses information stored by the change control tool and the planning tool. The output of this tool is used by the test manager to report test status. The output could be one graph or several, where each graph depicts one aspect of status.

The Test Quality tool

The purpose of the Test Quality tool is to produce a report on test quality in terms of coverage. This tool accesses the System Definition tool and the Planning tool and develops, in tabular form, a coverage report as seen from several views. The highest-level report is coverage by test area. The tool retrieves all the testcases that address the test area and provides percentages of testcases available to testcases run, working, being tested, etc. Similar reports are generated for all the system elements in the system definition list i.e., transactions, commands, databases. The reports generated by the tool are used by the test manager to ensure that all elements of the system are being tested and to determine how well the testing is progressing.

The One-page Report tool

The One-page Report tool produces a single-page report from all the current data that provides a summary of the test viewpoint of the current status of testing. This tool must gather information and present it in a concise format using only well-documented algorithms to summarize data or predictions. This tool is valuable if everyone accepts the output as valuable and can cause heated discussion if the output is not accepted. The source data for this output report is gathered as follows:

IRs	Change Content
SMRs	Change Content
Testcase	Planning tool
Coverage	Test Quality tool

The tool must report whether the release is on schedule, if the estimates for IRs seem correct, and if the current closure rate of SMRs is adequate. Calculations of the reported values use algorithms that are simple but must be developed over time and be accepted by all project management. For example, if SMRs have been closed at the rate of four a day on the average, then a straight line with slope 0.4 SMRs/day is an estimate of how many SMRs should be closed. If the actual number is not within 10 percent of this value, then the progress of SMR closure should be addressed. The test manager uses this report as a communication device and it is distributed to all the project's managers.

Summary

A complete set of integrated tools needed to automate testing includes tools for: content control, testcase maintenance, testcase execution, test metrics generation, and quality documentation.

The tools for content control must track requested changes and changes resulting from testing. The content tracking tools should also track a change through the building process so that the status of a change is available and current at any time.

Tools are needed for test maintenance and test development. These tools are analogous to the tools needed for supporting software development with the difference that these tools are usually keyed by testcase identifier.

The specific tools support configuration control, planning, specification generation, the compiler and its set up tools, prototyping support tools, cross-reference and coverage tools, and a generalized tool chest capability.

The test execution tools support not only the various methods of test execution, but must support testcase selection, testcase optimization, and testcase status and error reporting.

The test metrics generators must furnish metrics for discrepancy discovery and closure rates, SMR closure rate, feature testing status, testcase status, and quality of the software. The software quality must take into account the quality of the testing as well as the quality of the software. Many of the metrics should also be reported as a comparison to the same metric for past releases.

Appendix A

Testable items

I. Data Set Conditioning

    1. Valid records—enter "most" valid data combinations into the database. The reason, "most" is specified is that all record configurations would require a vast number (i.e., for IAM all would approximate 10,000 records).

    2. Invalid records—create invalid records to verify that invalid cases are handled properly. One possible source of invalid records is data that will generate the documented error messages. Special attention should be given to the following:

        a. Data that will fail validation checks. One method for creating these records is to enter data and then, if possible, change the validation table. Another method is to create records using a database access tool (e.g., the IBM IMS DLT0 tool) that allows invalid fields.

        b. Records that are database illogical. Create conditions that simulate database problems, e.g., use an old copy of one of several linked databases. This is an out-of-sequence condition that can be caused by refreshing one database, of an integrated set of databases, while others are left alone. If possible, enter duplicate data.

        c. Records that are application illogical. This causes conditions that the programs do not expect, e.g., a database might be constructed so that a 01 segment implies having a 05 segment. Using a database access tool, the 05 segment can be deleted without violating database logic but leaving the 01 segment is illogical as far as the application is concerned.

        d. Records that are not integrated. If the system's records are integrated, the tester can uncouple records by using a database access tool (e.g., DLT0).

e. Record with an invalid destination. Create a record with an invalid destination, such as a printer name that does not exist, to verify that the software rejects the message and does not deliver it to the bit bucket.

3. Every field supplied. Verify that the software correctly handles at least one record with every field containing completely specified data.

4. Minimum data. Create a record with the least amount of data to verify that the software can handle this condition.

5. Maximum data. Supply enough valid data to represent reasonable, userlike high-volume data to verify that the software can handle this many related records.

6. Multi-line. Supply enough data for one record to display on more than one line to verify that the software can display multi-lines.

7. Multi-page. Supply enough data for one entry to display on more than one page to verify that the software can display multi-pages.

8. Excess data. Valid data that will not be used because the input parameters used will exclude the data.

9. No data. Initialize the database with no data to verify that the software can handle an empty database.

10. Extra invalid records. Input records containing errors that would normally be reported but are not because the input parameters chosen eliminates processing of these records.

11. Constructed records. During normal processing, all records produced by an application are logical, i.e., the record can be processed without errors. However, illogical records can be produced by the application if errors occur during processing. A database access tool, e.g. DLT0, must be used to construct illogical records to simulate conditions that might exist.

Input

1. Valid input—use valid data to verify that the software handles the normal case. If more than one input option can be specified, the tester should verify that the software can handle at least the following valid cases: a single option, several options, and maximum options. Valid destinations for communication software should include as many valid destinations as the hardware configuration allows. It is mandatory to test each supported device but combinations such as three TTYs on one dial-up port might not be possible to test because of hardware limitations. These should be noted in the test specification.

2. Invalid input—use invalid data to verify that the software reacts reasonably. Include data with several variations if more than one field is validated. Also, note that some invalid conditions can be caused by the

incompatibility of valid options. Invalid destinations for communication software should include: invalid devices, database illegal names, devices not supported, legal names not recognized by the software, and null destinations.

3. No input—run with no input data entered to verify that the software handles this condition.

4. Maximum input—run with all possible data fields completely specified to verify that the software handles this condition.

5. Default values—if default values can be used, verify that the correct defaults are supplied and that the defaults can be overridden by supplying valid values when appropriate. Also, try invalid overrides to show they are rejected.

6. Hardware conditions—supply input for routing messages to hardware of various types as well as hardware with error conditions, which will be checked when the software executes. Conditions are: unit not available, unit quits in the middle of output, and unit quits during handshake.

7. Duplicate control card (batch)—run batch runs with duplicate control cards to verify that the software can handle this condition.

Execution

1. Terminal variations (on-line)—verify that software that is specified as capable of operating on various terminals is operational on each type.

2. Execute again—using the same input data, perform a transaction or run which changes the database again and verify that the software correctly handles duplicate requests (e.g., deletes do not find record, adds are rejected as duplicates, and updates are either accepted with no change or are rejected as no change). One possible user error to be guarded against with this testable item is performing batch runs twice.

3. Execute using special processing—execute batch runs using special process facilities. For the IAM system, the Reports Administrative Facility provides a special processing situation.

4. Hardware variations—if software can send messages to various hardware types, verify that the software correctly handles the normal conditions for each type of hardware and for each message type (if more than one). Include the following error conditions, all of which must be forced during execution:

   a. Unit not available. An error condition that must be tested is to verify that the software operates correctly when a unit is not available. There are two sub-cases. First, an alternate unit is available, and second, an alternate is not available (either it was not specified or it actually is not available).

b. Unit quits in middle of output. Verify that the software correctly handles the condition for distributing to a unit that quits during distribution. The tester can simulate this case by forcing the unit to quit even if this involves pulling the plug.

c. Unit quits during handshake. Verify that the software correctly handles the condition of the unit quitting while the handshake is going on. The tester can simulate this condition in a dial-up device by forcing the unit off-line while the answer-back sequence is going on.

5. Hardware paths—if the software communicates with other software in another system via various hardware paths, verify that each path is operational using a reasonable sample of messages.

6. Execute restart capability—if a batch run has a restart capability, such as the IBM checkpoint facility, verify that it executes correctly.

7. Buffer size variations—if software uses buffers, ensure that minimum, maximum, and a middle value is specified for buffer allocation size. A special case is the exact record size.

Result Verification

1. Check that output conforms to standards—for example, the IAM System on-line messages should be in the form: SYSNNNT Message Text, where SYS identifies the IAM system component system, NN is the assigned number of the message, T is the type with values (I = informational, E = Error, D = database inconsistency, and W = Warning), and that the message text is meaningful to the user. Some batch standards that the tester should verify might be:

a. Condition code usage should be standard (e.g., condition code = 000 should only indicate successful completion).

b. All messages to the user, especially error messages, should be understandable.

c. Ensure that correct IDR information is printed.

d. Verify the run documentation.

e. Verify that the run is step re-startable.

f. Check that the batch output corresponds to on-line format fields—the field names on the report should correspond to the on-line format field names that display the same data.

g. Check that the data layout is compatible with the report and query systems file representations. Also, compare the file representation and the source of the data set description to ensure that the data set descriptions are correct.

2. Check data usage via other software—for on-line formats, if the data is displayable on other formats, verify that the formats can access and display the data. Pay special attention to formats that use algorithms to select data to be displayed. For batch runs, cross-check batch results with other reports that contain the same information; otherwise, verify using on-line system.

3. Verify invalid data rejected—the tester should verify that invalid input was rejected and that attempts to use invalid data, entered as part of the database conditioning, were rejected. Using the on-line system, find some of the rejected records in the pre-conversion database and verify that the algorithm for conversion rejected these records. Note, the tester must understand the conversion algorithm.

4. Verify addition via tools—even though the execution of a transaction or run indicated that a record was added, an independent means should be used to verify the addition. For batch work, the on-line system or a data set print utility can be used. On-line checks could use a database access facility (e.g., DLT0). After creating a new record, check that it exists and its structure is correct. Special attention should be paid to seed records (initial records in a database). If the entire database is new, e.g., was converted from a previous database, pay close attention to the first and last records.

5. Verify data set changes via tools:

   a. Using a database access tool such as DLT0, dump records before and after an update to verify changes.

   b. Pay special attention to the first and last records in a database, as well as the records from the data set conditioning step, which contains all the specified fields and maximum data.

   c. Use the on-line system to verify updates if they can be viewed with on-line formats.

   d. Ensure that invalid records were not changed.

   e. Verify that extra data entered in the data set conditioning step remains in the database.

6. Verify delete via tools:

   a. Using a database access tool such as DLT0, verify that a deleted record is gone.

   b. Use the on-line system to verify that the record is gone.

7. Verify only requested data present. If input options can select subsets of data, verify that no unexpected data has been selected. Further verify that records which were supposed to be identified were. Previous results can be used as a comparison.

8. Check multi-page output—verify that the software correctly handles the boundary condition of displaying or printing multi-page output. Special

attention should be paid to the last record ending a page, the first record of the following page, and the last page.

9. Check multi-line output—verify that the software correctly handles the boundary condition of multi-line output. Special attention should be paid to records input during the data set conditioning step.

10. Check that receiving system accepts data—the tester must coordinate with the tester of the interfacing system and verify that the other system accepts and acts upon the data passed to it.

11. Check receiving system rejects invalid data—the tester must coordinate with the tester of the interfacing system and verify that the system rejects invalid data.

12. Check messages received—message distribution transactions must be able to communicate with various types of hardware/software combinations. The tester must verify that each type of hardware received the messages and that each possible path was utilized.

13. Verify new screen restrictions (on-line)—if a transaction results in screen changes, verify that the new screen restrictions are proper.

14. Verify subsequent runs use of data (batch)—verify that downstream runs that use the new data from prerequisite runs operate correctly. Pay special attention to runs that update the new data. Another special case is deleted data. Ensure that deleted data does not appear in subsequent runs. Often, data is marked for deletion rather than being physically deleted immediately.

15. Check run times (batch)—verify that the run time has not changed significantly from previous runs. If the run time is long, (expected to exceed two hours in actual use) the tester should consider requesting that restart capabilities be provided. Note that all programs that run in batch mode in the background in an on-line environment, usually generate backup and backout files.

16. Check control totals (batch)—verify that the expected number of records were processed. Previous results, or results from prerequisite runs, can be used to determine the correct number.

17. Check log information (batch)—if a run uses a log tape to report information, the tester should verify that the log tape information contains all the expected data.

18. Check report break-point (batch)—verify that the software correctly handles the boundary condition of putting out multiple reports. Special attention should be paid to the last record of the old report and first record of a new report.

19. Check results for 0, −, and + values—verify that the calculation algorithm correctly computes to produce negative, zero, and positive val-

ues. The tester must know the algorithm and will probably be required to condition the databases to produce the needed values.

20. Check tables for first and last entries—if the software uses tables, ensure that the first and last entries, as well as a few in the middle, are used correctly. This choice is a "standard" method of ensuring that a table is accessed correctly. Also, choose at least one entry that has the maximum number of characters so that this condition can be checked.

21. Verify sort—if a sort is used to pre-process the output, verify that the sort is performed correctly.

22. Verify date manipulations—some algorithms are date sensitive, in that, depending upon the date input, the output will be different. Verify that date manipulations are correct using test data that includes: last year, last month, current date, next month, next year, and years greater than 2000.

23. Verify translations—some algorithms are concerned with translating data from one form to another. Verify that the translation is performed correctly.

Appendix B

Checklists for testcase design

I. Checklist for add software
   1. Data set conditions
      a. valid records
      b. invalid records
      c. every field supplied
      d. minimum data
      e. excess data
      f. no data
      g. constructed records
   2. Input
      a. valid input
      b. invalid input
      c. no input
      d. maximum input
      e. duplicate control card
   3. Execution
      a. terminal variations
      b. execute again
      c. execute using special processing
      d. execute restart capability
      e. buffer size variations
   4. Result verification
      a. check output conforms to standards
      b. check data usage via other software
      c. verify invalid data rejected
      d. verify addition via tools

       e.   verify only requested data present

       f.   verify subsequent runs use of data

       g. check run times

       h. check control totals

       i.   check log information

II. Checklist for communication software

   1. Data set conditions

      a. valid records

      b. invalid records

      c. multi-line

      d. multi-page

      e. excess data

      f.   no data

   2. Input

      a. valid input

      b. invalid input

      c. no input

      d. default values

      e. hardware conditions

   3. Execution

      a. terminal variations

      b. execute again

      c. execute using special processing

      d. hardware variations

      e. hardware paths

      f.   execute restart capability

      g. buffer size variations

   4. Result verification

      a. check output conforms to standards

      b. verify invalid data rejected

      c. verify change via tools

      d. check multi-page output

      e. check messages received

      f.   verify subsequent runs use of data

      g. check run times

      h. check control totals

      i.   check log information

III. Checklist for delete software

   1. Data set conditions

      a. valid records

      b. invalid records

      c. every field supplied

      d. minimum data

   2. Input

      a. valid input

   b. invalid input
   c. no input
   d. maximum input
   e. default values
   f. duplicate control card
  3. Execution
   a. terminal variations
   b. execute again
   c. execute using special processing
   d. execute restart capability
  4. Result verification
   a. check output conforms to standards
   b. check data usage via other software
   c. verify invalid data rejected
   d. verify delete via tools
   e. check multi-page output
   f. verify subsequent runs use of data
   g. check run times
   h. check control totals
   i. check log information
IV. Checklist for interface software
  1. Data set conditions
   a. valid records
   b. invalid records
   c. every field supplied
   d. maximum data
   e. multi-page
   f. excess data
   g. no data
   h. constructed records
  2. Input
   a. valid input
   b. invalid input
   c. no input
   d. default values
   e. duplicate control card
  3. Execution
   a. terminal variations
   b. execute again
   c. execute using special processing
   d. hardware paths
   e. execute restart capability
   f. buffer size variations
  4. Result verification
   a. check output conforms to standards

    b. verify change via tools
    c. verify only requested data present
    d. check receiving system accepts data
    e. check receiving system rejects invalid data
    f. verify subsequent runs use of data
    g. check run times
    h. check control totals
    i. check log information
    j. verify sort
    k. verify data manipulations
    l. verify translations

V. Checklist for query software

   1. Data set conditions
      a. valid records
      b. invalid records
      c. every field supplied
      d. minimum data
      e. maximum data
      f. multi-line
      g. multi-page
      h. no data
      i. extra invalid records

   2. Input
      a. valid input
      b. invalid input
      c. no input
      d. maximum input
      e. default values
      f. duplicate control card

   3. Execution
      a. terminal variations
      b. execute again
      c. execute using special processing
      d. execute restart capability

   4. Result verification
      a. check output conforms to standards
      b. check data usage via other software
      c. verify invalid data rejected
      d. verify only requested data present
      e. check multi-page output
      f. check multi-line output
      g. check run times
      h. check control totals
      i. check log information
      j. check report break-points

   k. check for 0, − , + values
   l. check tables 1st, last entries
   m. verify sort
   n. verify data manipulations
   o. verify translations
VI. Checklist for state-dependent software
   1. Input
      a. valid input
      b. invalid input
      c. no input
      d. maximum input
      e. default values
   2. Execution
      a. terminal variations
      b. execute again
   3. Result verification
      a. check output conforms to standards
      b. verify invalid data rejected
      c. verify new screen restrictions
VII. Checklist for update software
   1. Data set conditions
      a. valid records
      b. invalid records
      c. every field supplied
      d. minimum data
      e. maximum data
      f. multi-line
      g. multi-page
      h. no data
   2. Input
      a. valid input
      b. invalid input
      c. no input
      d. maximum input
      e. default values
      f. duplicate control card
   3. Execution
      a. terminal variations
      b. execute again
      c. execute using special processing
      d. execute restart capability
   4. Result verification
      a. check output conforms to standards
      b. check data usage via other software
      c. verify invalid data rejected

    d. verify change via tools
    e. check multi-page output
    f. verify subsequent runs use of data
    g. check run times
    h. check control totals
    i. check log information
    j. check for $0, -, +$ values
    k. check tables 1st, last entries
    l. verify sort
    m. verify data manipulations
    n. verify translations

Appendix C

Checklist summary

Checklist Summary

		Software Function						
TEST DATA SET	AREA	Testable Items	ADD	COMM	DELE	INTF	QURY	STDP	UPDT
	CONDITIONS	Valid records	H	H	H	H	H		H
		Invalid records	H	H	H	H	H		H
		Every field supplied	H		M	M	H		H
		Minimum data	M		M		H		H
		Maximum data				H	H		H
		Multi-line		H			H		H
		Multi-page		H		H	H		H
		Excess data	M	M		M			
		No data	L	M		L	L		M
		Extra invalid records					H		
		Constructed records	M			H			

Checklist Summary

	Software Function						
Testable Items	A D D	C O M M	D E L E	I N T F	Q U R Y	S T D P	U P D T
INPUT							
Valid input	H	H	H	H	H	H	H
Invalid input	H	H	H	H	H	M	H
No input	M	M	L	M	M	M	M
Maximum input	H		H		H	H	H
Default values		H	M	M	M	L	M
Hardware conditions		H					
Duplicate control card	H		H	H	H		H
EXECUTE							
Terminal variations	H	H	H	H	H	H	H
Execute again	H	H	H	H	H	H	H
Execute using special processing	H	H	H	H	H		H
Hardware variations		H					
Hardware paths		H		H			
Execute restart capability	H	H	H	H	H		H
Buffer size variation	H	H		H			
RESULT VERIFICATION							
Check output conforms to standards	H	H	H	H	H	H	H
Check data usage via other software	H		M		H		H
Verify invalid data rejected	H	H	H		H	M	H
Verify addition via tools	H						
Verify change via tools			H	H			H

Checklist Summary

Testable Items	Software Function						
	ADD	COMM	DELE	INTF	QURY	STDP	UPDT
Verify delete via tools				H			
Verify only requested data present	H			H	H		
Check multi-page output		M	M		M		M
Check multi-line output					M		
Check receiving system accepts data				H			
Check receiving system rejects invalid data				H			
Check messages received		H					
Verify new screen restrictions						H	
Verify subsequent runs use of data	H	H	M	H			H
Check run times	M	M	M	M	M		M
Check control totals	M	M	M	M	M		M
Check log information	L	L	M	L	M		L
Check report break-points					M		
Check for 0, −, + values					H		H
Check tables 1st, last entries					H		H
Verify sort				H	H		H
Verify data manipulations				H	H		H
Verify translations				H	H		H

RESULT VERIFICATION

Appendix D

Automated test tool specifications

Four tools are discussed in this section: a content control tool, a testcase execution tool, a testcase quality tool, and a tool to produce a one-page report.

Content control

These are the testers' specifications for a software change tracking system called Change Request Tracking System (CRTS) [BOND90]. To be a useful standard tool, CRTS must support projects of various sizes and in various environments. The tester's major responsibility is to determine the quality of the software under test. However, testers have many other responsibilities depending on the project. Therefore, CRTS must be capable of supporting a broad spectrum of current and future tester responsibilities. The requirements from the tester's point of view are divided into several sections, which address: data content, interfaces, CRTS administration, system flexibility, security, reporting, performance, a help function, and the CR flow.

The term *Change Request (CR)* is used to represent an item tracked by CRTS. This includes what was previously called SMRs, SMR items, and IRs. CR is introduced to unite all the sources of software change into one system. Using one system to track all software changes is more efficient because it eliminates interfaces, and if the system is used throughout the company, then no training is required when someone transfers to a new organization. While CRs can be presented in different formats to different classes of users (e.g., developers, testers, etc.), only one record structure is used. These specifications define the origin of the data needed by testers.

Data content and validation

It is necessary to specify and understand the data content requirements before functionality can be specified. Each of the following entries specifies the field name, its contents, and the applicable validation rule for CR input. Validation rules are under the control of an administrator program that is discussed in the CRTS Administration section. Certain validation types are optional, and will be stored in a database that can be updated by an administrator program. The fields listed are defined as either character (c) or numeric (i) and have a specified length. A repeated field, with a specific number of repetitions, is indicated by an integer and an "*". If the number of repetitions is indeterminate, the character *n* is used before the *. Note that the repeated fields could be implemented as linked lists. Validations should adhere to these restrictions.

1. CRID FIELD (change request id) (c10)
   This field must have the following:

       PPCCYYYYJJJXXX

   PP is the project identifier. CC is one of a specified set of valid values determined from the user's login that indicates the source of the CR (customer, development, requirements, etc.). This value should be overridden if a tester is entering a customer CR from their login. YYYY is the year. JJJ is the Julian date, and XXX is the sequence number, which starts at 1 each day and increments by 1 to the number of CRs input during the day. This field is automatically generated for each CR entered and, therefore, requires no validation.

2. CR SEQUENCE FIELD (i4)
   This field can be used to determine the total CRs for a release of each product and is set by the administrator. It is a numeric field that is automatically generated, incremented, and reset based on a new release.

3. SUBSYSTEM (c20)
   The subsystem in which the trouble was identified. The default validation on this field is strictly a length validation of 20 characters. Optionally, values could be restricted to a specified set of product-specific areas, with a table of valid areas, that could be updated by testers.

4. SOFTWARE RELEASE VERSION IDENTIFIER FIELDS (3*c10)
   Three fields are required for the product release versions:
   - The release of the software in which the problem was discovered.
   - The release in which the problem is scheduled to be fixed.
   - The release in which the problem actually was fixed.

       All fields are validated against a table of valid release identifiers specified by an administrator.

5. STATUS FIELD (c15)
   This field shows the current status of the CR and is initialized automatically and updated automatically or manually. The values are checked against a table of possible entries that can be altered by the administrator. (See sample table TABLE D-1.)

Table D-1. CR Status.

Name	Meaning
NEW	Initial status for field CR
REQ	In requirements
DEV	In development
RDT	Ready to be delivered to test
ADT	Tester approved for delivery to test
DTF	Delivery to test failed
INT	In test
FT	Test failed, returned to development
PUT	Passed unit test
PMUT	Passed multi-unit test
PPT	Passed product test
PIPT	Passed inter-product test
PET	Passed evaluation test
CL-XXX	Closed with reason
	XXX = DEF deferred
	XXX = FPC fixed by a developer in a previous change
	XXX = REJ rejected
	XXX = CAN canceled
	XXX = TF problem was fixed by tester
	XXX = ROL rolled to future release

6. TROUBLE AREA (c20)

   This field includes the module, transaction, procedure, or document in which the trouble was discovered. It has the same validations as SUB-SYSTEM but additional values might be required.

7. PERSON ENTERING CR (c10)

   This field should be pre-populated based on login ID and CC characters of the CRID field but should be overridable. This must be checked against a list of valid users.

8. ABSTRACT (c80)

   This field contains a short description of the problem. Validation is based on the length for maximum size (c80) and minimum size (c10).

9. HOW FOUND FLAG (tester input) (c10)

   The how found flag value Is checked against a default list of possible choices shown in TABLE D-2.

10. HOW FOUND TEXT (tester input) (c100)

    Testers use this field to describe conditions associated with how the problem was found so that the test can be duplicated.

11. SEVERITY (c1)

    This field is validated against a default list of severities as follows:

    1 Critical—the system is down or a critical interface between two systems has failed.

Table D-2. Method of Discovery.

Name	Meaning
TCXXXXXXXX	Found by the testcase named XXXXXXXX
FEATXXXXX	Found in testing feature XXXXX
REG	Found in regression testing
EXER	Found while exercising the software
etc.	

  2 Severe—the problem is inhibiting an important area of the system from functioning and there is no acceptable means of circumvention.

  3 Functional—these are functional problems that impact users, system administrators, or maintenance personnel. Acceptable workarounds might exist or the software recovers by itself but the problem cannot be deferred indefinitely.

  4 Inconvenient—these are minor or noncritical problems with the system or component.

12. PRIORITY (tester CR) (c1)

This field requires two validations. First, the field is reserved only for a tester. If the PERSON ENTERING CR is a tester, then this field can only be one of the entries shown in TABLE D-3.

Table D-3. CR Priority.

Name	Meaning
A	Need ASAP
B	Need for this release, but not immediately
C	Can wait for a later release

13. ASSIGNMENT FIELDS (n*c10)

Based on CRID information and some system tables, the system automatically assigns the Tester (c10), Developer (c10), Documenter (c10) and Analyst (c10) responsible for this SUBSYSTEM.

14. ENVIRONMENT FOUND (n*c10)

On what machine was the CR found? Validation is based on a table with all existing valid machine names. If no table exists, a simple field length validation is all that is required. Multiple machine names can be entered.

15. ENVIRONMENT FOR FIX (n*c10)

On what machine should the problem be fixed? Validation is the same as ENVIRONMENT FOUND. Multiple machine names can be entered.

16. CHANGE REASON (developer supplied field) (c15)

Validated against a list of reasons for change. These are stored in a table maintained by the administrator and tunable by project. A sample of the defaults provided are shown in TABLE D-4.

Table D-4. Change Reason.

Name	Meaning
Software	Problem isolated to the code
Feature	Need a new feature to resolve this item
Requirements	Problem in requirements
Documentation	Documentation and software do not agree
Missing Info	Did not have information when needed
Human Eng.	Not presented well
Spelling	Spelling error made
No Change	Procedural error
etc.	

17. DATE FIELDS (n*c10)

Validation is either the MM/DD/YYYY or YYYY/MM/DD date validation.

- date discovered—the default entry should be the current date and the default should be pre-populated when the CR entry screen is initially displayed.
- date CR entered also defaults to the current date.
- date development is complete.
- date last updated.
- date the five test phases—unit, multi-unit, product, inter-product, and evaluation—are completed.
- etc.

18. DOCUMENTATION REQUIRED? (c1)

Is a documentation change required? A one-character required field with two valid responses, 'y' or 'n'.

19. EXPLANATION FIELDS (c1000)

This is a text area where no validation is required. The area should be a window that can be paged back and forth. Each entry has a date and the writer's initials. The information includes:

- problem description,
- test results,
- resolution, and
- why delivery failed.

20. SOFTWARE DELIVERY FIELDS

Software delivery fields can be passed via an interface to a build system and provide instructions for delivery of software, such as:

- Software to be moved.
- Identify source code that was changed.
- Configuration control level of software to be used.
- Size of software.
- Compilation date.
- Deliver to which machine.
- Where to find software if not in a default data set.
- Special instructions.

21. COPY TO LIST (n*25)

The capability should exist to generate the list automatically. A default table should also be available for each subsystem and could be used to determine this list. A manual entry added to this list means that for any change in the CR's status, the person added will also get a copy of the updated CR. Because anyone should be able to receive a copy of the CR, electronically, names can be added to the default list. Names can also be deleted from the default list. No validation is required for the list.

22. PERSON WHO LAST UPDATED (c10)

Same validations as PERSON ENTERING CR.

23. RELATED TO CRs (n*c10)

Each CR entered should be verified as a valid, existing CR number. Multiple CR numbers can be entered.

24. PRICING (c10)

Cost to fix problem. No validation other than field size.

25. COST OF IMPACT (c10)

Cost of work-around. No validation other than field size.

26. DESCRIPTION OF WORK AROUND (c200)

A description of the actions that can be taken to avoid the problem.

Interfaces

This section covers both CRTS automated interfaces and human interfaces.

Automated interfaces   CRTS must be able to interact with the software build system, i.e., BUSY. As the build system processes the software, the software status in the CR must be changed to reflect the current status.

Human interfaces   CRTS should be user-friendly and should use current technology. The human interface used to operate CRTS should be consistent with the human interface for the products whose changes it tracks. The user should have access to all levels of data associated with a CR without needing to step through menus. This includes the capability to view validation "rules" via a simple auxiliary menu. Other parts of a CR or other CRs should be accessible. CRTS should be accessible from any environment. CRTS users should be known by their login (not by their terminal ID).

Each user will be put into a user class. Users can then perform functions that are associated with their user class. A representative sample of the classes are:

- administrator
- coordinator
- scheduler
- developer
- user
- tester
- other

A user class can be authorized to perform any or several of these functions.

CRTS administration

The system must contain a full set of operational utilities to enable (with proper controls and restrictions) the capability of deleting a record, changing any field in a record (including the key), or adding a record manually (without validations). A complete set of backup, restore, compression (if necessary), and archival facilities must also be provided. The archive must be accessible by the retrieval procedures if requested by the user but should not be accessed during normal processing.

The rest of this section describes an Administrator Program (AP) that maintains all the tables in the system. The AP should be user-friendly! Sometimes, the AP might have to search existing records and make appropriate changes when a table is changed. A few examples of when such changes are necessary are:

- When a developer (or tester, etc.) leaves, all CRs assigned to that developer (or tester, etc.) should be changed to one or more people.
- When a table, like the permission table (security section), is changed, all records containing user permissions in that category or in a new or deleted category must be changed.

The AP should produce reports and/or on-line messages that show the changes made. Changes that cannot be supported should be reflected in the reported messages. This allows an administrator to delete and re-add the record, which should be an on-line interactive process.

System flexibility

The change request tracking system needs to have the flexibility for user customization at the project level. Specifically, this involves overridable default field validations and pre-populations (deriving fields from values in other fields). In addition, any status changes, explicit or implicit, must be controllable by the user to the point of elimination.

The implication of these requirements is that the delivered system must contain default values for all user-modifiable parameters. However, the user should then be able to easily and effectively change those values, even to the extent of being able to invoke a user-provided process or program on a project basis. There must be proper controls and restrictions associated with the ability to change the customizable parameters.

Security

The security specification discusses permissions, changing permissions, and an administrator program to change permissions. Authorization to enter or change CRs is required whenever the system is entered. The classes of CR system users that would need authorization permission defined would include:

- Hotline personnel
- System testers
- Software developers
- System administrators
- System engineers
- Management
- User (field or internal)
- Guest testers
- Component (unit) testers
- Software build people
- Other (external interfaces)

The users would be assigned, on a project basis, one of the following authorization classes:

- Add, Update, Delete, or Close all CRs entered for a particular project or module. This is analogous to a "super-user."
- Perform only a prescribed number of operations in the CR system. This is done using a table set up by the administrator. Permissions to do a particular function is turned on for each user. A suggested default setup is shown in TABLE D-5.

Table D-5. User Permission.

Category	Add	Update	Delete	Close
Tester	Y	Y	Y	Y
Developer	R	Y	N	N
Manager	N	N	N	N
User	Y	Y	N	Y

The values are defined as Y = user can perform function, N = user cannot perform function, and R = restricted usage, meaning that a validation routine is supplied to determine Y/N. Note that all fields are stored in a database and can be modified for each product for their own needs. Permissions should be kept in a table that can be easily viewed and updated by the CR system administrator. The table must also be easily expanded to include more functions, including retrieving reports, receiving mail, and other statuses.

Reporting

The system will not be considered complete without a customizable set of reports. These reports must have enough diversity to enable selection of specific or all records for either detailed or statistical summary reports.

In considering the requirements for a new tracking system, attention must be focused on generating reports. From experience, you will know the types of reports that will be required for various user communities. The rest of this section discusses general reporting requirements and specific reporting needs within each category. The user communities fall into three categories:

1. Testers.
2. Developers and requirements writers.
3. Project management and users.

General reporting requirements

General reporting requirements apply to all three categories. They are:

- The ability to specify formatted reports using a report-writing language and save the report requests for later execution.
- Language capabilities that include:
  a. fields to be reported.
  b. selection criteria (logical operators and wild cards)
  c. sorting criteria.
- The ability to input selection criteria via a named file.

- The ability to generate an unformatted report quickly.
- The ability to direct output to an on-line screen, batch file, local printer, or high-speed printer.
- The ability to support graphic output.

Specific test reports

The specific test reports that are generated so testers can monitor progress of a release are:

- The report for the individual tester. This should be on a request basis and only generate output for that user.
- A report, in matrix form, for management showing the tester on the left and a choice of fields across the top, such as the number of items to test, number of items completed, etc.
- Ability to generate a report of CRs based on some time range.

Specific developer reports

The specific developer reports that are generated for development groups to monitor progress of a release are:

- The report for individual developers. This should be on a request basis and only generate output for that user.
- A report, in matrix form, for management showing the developer on the left and a choice of fields on the top of the report.
- The ability to generate a report of CRs based on some time range.

Specific user reports

There are also specific user reports that include:

- A report of all CRs that are resolved by a particular release of the application software. This should be flexible enough to support the sorting requirements of the user.
- A report of all CRs that have been closed since their last report.
- A report for a particular user.
- The ability to reformat the output report to meet the users' needs. This includes the ability to change the field values that the report is sorted on as well as the ability to change the overall appearance of the report.

Performance

The CRTS system must support multiple users of any type. Any action on one CR should process within five seconds under an average load. Actions on multiple CRs are considered to be reports and do not process within five seconds. Reports can be run on-line or off-line at the user's discretion. Report response time is dependent on the type and data content and cannot be specified, but should be reasonable.

Help function

The system should contain a generalized HELP function. HELP should be interactive and should consist of two levels. If HELP processing is started from a field (cursor based), then the first level is the valid field values with names (no description), and the second level should provide a description of each value. If a value is selected, at the first or second level, then when HELP is ended, that value is entered into the field from which HELP processing was started. If HELP processing is started with the cursor in the home position of the screen, the HELP should describe the screen usage.

Change request flow

Figure D-1 illustrates the flow of a CR through CRTS. A CR can be entered by a requirements person for an enhancement, by a tester, a developer, the user, or a support center acting for the user. The CR is then scheduled for a release. This typically involves a cooperative effort among various project personnel, including developers, testers, project managers, etc.

When the CR is being addressed by development, several paths can be followed:

1. The software is changed and the CR is ready to be included in a release. Then the CR is typically "sent" to the build process.
2. The developer needs new requirements and the CR goes to requirements. After requirements are available, the CR is returned to the developer.
3. The change is more complicated than originally estimated and must be rescheduled for a future release.
4. The developer determines that no change is required and the CR is "sent" for no change processing.

No change processing determines if the CR can be closed or not. A CR does not require a software change if: the problem is already fixed, it is a data problem, it is an environment problem, or it is non-reproducible, etc. In these cases, the CR is closed. If no change processing determines that the CR should not be closed, it is routed to the appropriate personnel, determined by project guidelines, for further processing. The CR might then be returned to be scheduled, returned to developers, or returned to requirements writers.

After the software has been changed, it is built. The turnover to system test is controlled by system test accepting the changed software into their environment.

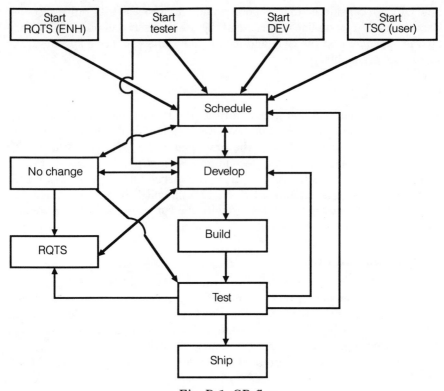

Fig. D-1. CR flow.

When the software (and the CR) is being tested, several paths could be followed:

1. The tester determines that the changed software operates as needed and the CR is passed.
2. The tester determines that new requirements are needed and the CR goes to requirements personnel. The CR then goes to development and build before coming back to the tester.
3. The tester determines that the changed software does not operate as needed and fails the CR. The CR might go back to the developer or a new CR might be opened.

CRs that are passed are ready to ship.

Testcase execution

Testcase execution specifies the Test Execution Tool (TET). The three components of TET are: the executive, the drivers, and the reporter. The functions of these three components are described but the specification concentrates on the interfaces between the components [BRUH88].

Component functionality

The components of TET are the executive, driver, and reporter. Executive controls the tests to be run by what driver and when. Driver communicates with the system being tested by applying stimuli, observing responses, and comparing actual responses with expected responses. It also provides the capability to create and maintain testcases. Reporter analyzes the results of the test execution and produces reports.

Executive

The basic features of the executive are:

- Controlling the drivers in a primary/secondary relationship.
- Controlling the reporter in a primary/secondary relationship.
- Scheduling the execution of test sections on one or more drivers.
- Interfacing with the tester/user.

Reporter

The basic features of the reporter are:

- Providing a report describing the test execution history from the start of the test period.
- Producing a report about a single test or test section.
- Producing a report about a group of tests.
- Maintaining the necessary history files for producing the preceding reports.
- Producing a Structured Query Language (SQL)-like arbitrary report based on the history file.

Driver

The basic features of a driver are:

- Activating and controlling several instances of a specific driver or different drivers by the executive at one time.
- Executing tests or test sections as specified by the executive.
- Retrieving the test or test section before executing it.

Figure D-2 shows the components of TET and the interfaces.

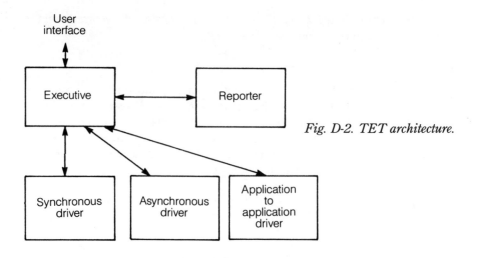

Fig. D-2. TET architecture.

Interface message specification

Messages passed between TET components are commands that the receiving component must act on, acknowledgements to messages, or information needed by the receiving component.

Message format

Messages should be passed in a packet format with each packet containing one or more aggregates. A standard syntax must be established so that new components can be easily added.

Message types

The messages must be one of the three types:

1. Request—a command sent from the primary component to a secondary component.
2. Solicited response—a status message sent from a secondary component to the primary component when requested.
3. Unsolicited response—a status message sent from a secondary component to the primary component, when necessary, to inform the primary of the secondary's status.

Protocol

Two interface components always exist in a primary/secondary relationship. The primary component can only send request-type messages to a secondary component. The secondary component must immediately respond with a solicited response-type message. A secondary can also send unsolicited response-type messages to the primary to provide current status information.

Every request is answered with either a positive acknowledgment response (pack) that the request was received, a negative acknowledgment (nack) response that the message was received but cannot be acted on, or a status response message that is used in other cases.

Message aggregates

The following message aggregates are used:

Control	Each message begins with a control aggregate. This aggregate contains the controlling information about the message e.g., solicited or unsolicited flag, command, name of sending component, and name of receiving component.
Driver status	Contains the execution status information for the driver. It is used to pass information to the driver and status back to the executive.
Test ID	Is part of any message about a test or test section and identifies the specific test and its origin.
Test status	Provides the status information about a test being executed by a driver.
Reporter status	Contains information to control the reporter and to pass reporter status back to the executive.

Executive-driver interface specification

The messages across the executive-driver interface specification can be related to either status or execution.

Status messages   The executive must know the current status of the drivers in order to schedule. The executive needs to be able to change the status of a driver, e.g., pause execution, resume execution. The driver will always respond to messages with either a pack or nack. The specific messages are:

- *Request Driver Status*. The executive uses a request driver status message to obtain the current status of a driver. The drive must respond with a driver status message.

- *Change Driver Status*. The executive uses a change driver status message to change the status of the driver to start, halt, pause, or continue.

- *Driver Status*. Normal letters send unsolicited driver status messages to the executive describing their status changes. When a test execution has completed, the driver notifies the executive using a driver status message.

Execution messages   The executive controls the execution of tests using execution messages to which the driver must respond with either a pack or a nack. The specific message is "Request Test Execution." The executive initiates and controls test execution using a request test execution message to start, pause, or halt.

Executive-reporter interface specification

There are two messages sent across the executive-reporter interface specification:

1. *Request Results Processing.* When a test or test section terminates, the executive sends the results to the reporter for processing.
2. *Request Action.* When the user requests a report, the executive relays the request to the reporter using a request action message.

Testcase quality

The purpose of QUEST, the QUality Evaluation SYstem, is to provide: a qualitative view of product test, an up-to-date status of test activities, and tailored management reports.

The process of testing software usually starts with each new piece of software being analyzed and an appropriate test plan being developed. A missing ingredient in this process is the carryover of experience from one test plan to the next. One method used to furnish instant experience is a checklist.

QUEST specifications

The checklists described in chapter 5 and shown in appendix A, B, and C are the essential reference data for a system that performs four processes:

1. Create testcases.
2. Update testcases.
3. Select testcases for execution on TET.
4. Report on status of selected testcases.

Figure D-3 shows the QUEST system architecture, which is discussed at a functional level. Further details of each of the individual components and processes are discussed in separate sections later. A functional overview of the system architecture is:

It is assumed that the Checklist Reference Database was established by some initialization process. The Checklist Reference Database contains entries that correspond to the summary checklist (appendix C). It is also assumed that this database can be permanently updated or changed by some utility process that is outside the scope of this book.

The create process interactively generates a Test Reference Database record for each software entity to be tested (FIG. D-4). The Test Reference Database is keyed by the tested entity identification (e.g., runid, transaction,

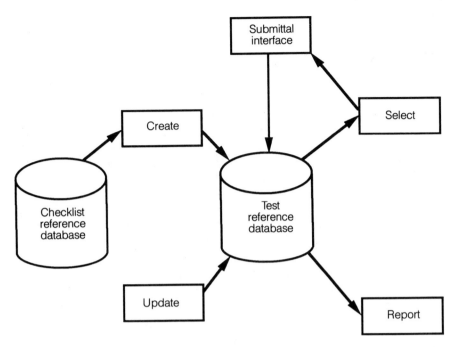

Fig. D-3. QUEST architecture.

interface, etc.), and contains items to be tested (ITBTs), testcases needed to verify the ITBTs, and the priority associated with the ITBT.

The update process operates on the Test Reference Database. This process updates records to indicate optional grouping information or to indicate the release for which a specific entity is to be tested.

The select process that is shown in FIG. D-5, selects the requested subset of testcases. These selections are passed via the submittal interface to TET. The user must define the selection criteria. The process automatically updates fields of each testcase selected to reflect its current status.

The report process that is shown in FIG. D-6, selects the requested subset of testcases and reports their status. The TET reporter reports testcase execution status while the QUEST report process reports contain testcase development status and testcase coverage information. The process requires the user to define the selection criteria and the output type.

The function of the submittal interface component of QUEST is to provide isolation between QUEST and execution tools. The submittal interface will receive requests and format them as needed by the batch or on-line execution tools and send the requests to the appropriate destination. The interface will receive test status returns and set the status in the Test Reference Database accordingly.

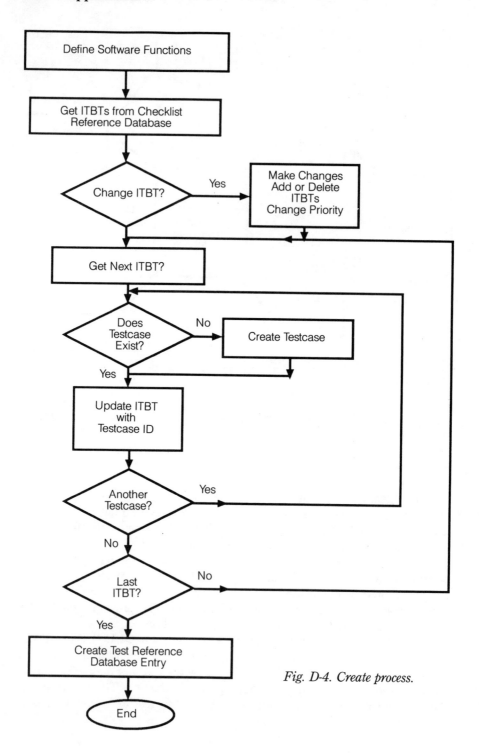

Fig. D-4. Create process.

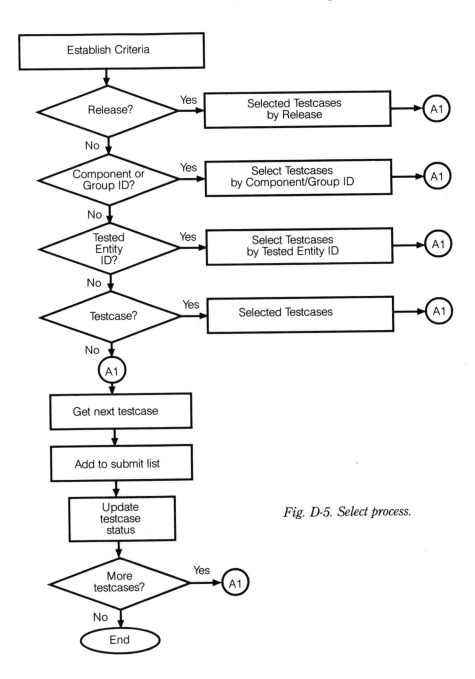

Fig. D-5. Select process.

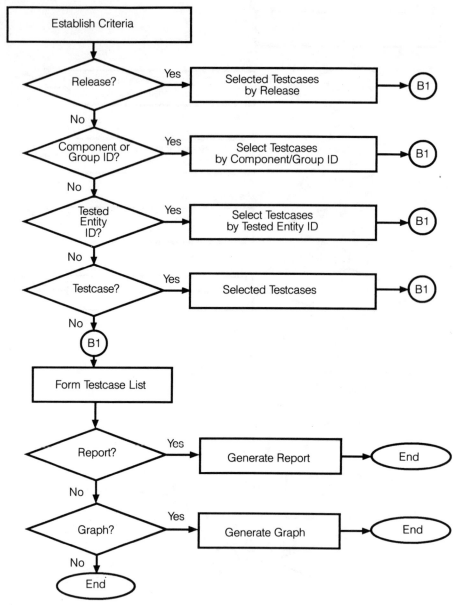

Fig. D-6. Report process.

QUEST process—CREATE

The create process that creates test reference database records is shown in
FIG. D-4. A step-by-step explanation of the create process is:

1. Define the software functions for the Software Under Test (SUT). One or more of the functions (column) from the checklist summary.
2. Get a list of Items to be tested (ITBTs) from the Checklist Reference Database. The ITBTs are the rows of the summary checklist for one column. Because software can perform more than one function, the list might have to have ITBTs from more than one column. Duplicate ITBTs are eliminated by the process without any intervention.
3. Allow tester to make changes as an interactive step for the tester to add or delete ITBTs or to change the priority of an ITBT. The tester can tailor ITBTs to the specific application's needs for a specific release. Note that permanent changes are made via an off-line utility process.
4. Get the next ITBT from the list developed in steps 2 and 3.
5. If a testcase exists to verify this ITBT, proceed to the next step. If a testcase does not yet exist to verify an ITBT, then the tester must create the testcase identification.
6. Associate the testcase with the ITBTs that the testcase verifies by updating the ITBT entry with the testcase identification.
7. Are more testcases necessary to verify this ITBT? If so, go back to Step 5. If not, proceed to Step 8. (Note: an ITBT might require several testcases.)
8. If this was not the last ITBT, then go back to Step 4.
9. If this was the last ITBT from Steps 2 and 3, then create test reference database entries and the process is finished.

QUEST process—UPDATE

The steps for the update process for amending the testcase data in the test reference database record is:

1. Find an individual record in the test reference database.
2. Update record to include:
   a. New releases for which entity will be tested.
   b. Optional subsystem grouping.
   c. New testcase used for testing entity to be tested, and etc.

QUEST Process—SELECT

The select process that is shown in FIG. D-5, selects the requested subset of testcases and passes them to the submittal interface for execution. The user must define the selection criteria. The process also automatically updates fields of the test reference database for each testcase selected to reflect its selected status.

1. The user must indicate the search criteria via input.
2. If the criteria is a release number, then select from the test reference

database all testcase segments that contain the specified release number. Form a list of these testcases. Go to step 6.

3. If the criteria is a component or group ID, then select all records from the test reference database that contain the specified component or group ID. Form a list of these testcases. Go to step 6.

4. If the criteria is tested entity ID, then select all records from the test reference database under the tested entity ID. Form a list of these testcases. Go to step 6.

5. If the criteria is testcase, then select all records from the test reference database that contain the specified testcase. Form a list of these testcases. Go to step 6.

6. Get the next testcase from the list established in steps 2, 3, 4, or 5.

7. Include the testcase on the list to be sent to the submittal interface.

8. Update the status fields of the testcase to reflect selection and update the parent statistics segment fields of the testcase to reflect active status.

9. Any more testcases? If yes, go to step 6.

10. If no more testcases, then the selection process is completed.

QUEST process—REPORT

The report process shown in FIG. D-6 selects the requested subset of testcases and reports their status. The process requires the user to define the selection criteria and the output type in the following steps.

1. The user must indicate the search criteria via input.

2. If the criteria is a release number, then select from the test reference database all testcase segments that contain the specified release number. Go to step 6.

3. If the criteria is a component or group ID, then select all records from the test reference database that contain the specified component or group ID. Go to step 6.

4. If the criteria is tested entity ID, then select all records from the test reference database under the tested entity ID. Go to step 6.

5. If the criteria is testcase, then select all testcase segments from the test reference database that contain the specified testcase. Go to step 6.

6. Form a list of the testcases selected in step 2, 3, 4, or 5.

7. Is output type Report? If so, generate the report and the report process is complete.

8. Is output type Graph? If so, generate the graph and the report process is complete.

9. Because no other output types are supported, the process is completed.

Note that enhancements to the system could include further output types.

QUEST component—submittal interface

The submittal interface provides all other QUEST components with an unchanging interface to TET. Changes to the input of TET or changes to the output of these tools only affects the submittal interface.

After being invoked by the select process, the interface initiates the execution of scripts or batch testcases and changes the status to reflect those submitted. The interface accepts notification, analyzes the returned data, and updates the appropriate status field in the Test Reference Database.

QUEST component—checklist reference database

The contents of the checklist reference database records are specified as:

- The software function [key]
- Testable items, test area, and number (e.g., A2)
- Priority (R—required, S—should be run, O—optional)

QUEST component—test reference database

The entries in a test reference database record include:

1. Test entity ID (Run ID, transaction, interface, etc.).
2. Type.
3. Component ID.
4. Group ID.
5. Usage priority (category).
6. Tester.
7. Release under test.
8. Number of items passed (by category).
9. Number of items active (by category).
10. Number of items failed (by category).
11. Number of items no activity (by category).
12. SMR.
13. Component subsystem of SMR.
14. Notes.
15. Checklist item (pointer to checklist reference database) [key].
16. Test category.
17. Description (pointer to checklist reference database).
18. Testcase counter.
19. Status.
20. Tested entity ID (run ID, transaction, interface, etc.).

One-page report

The One-Page status Report (OPR) provides a comprehensive system test status for the release under test. OPR provides management with release information and accurate estimations that allow them to make quality decisions. The information is organized into six sections with varying levels of detail. This specification defines the report content and the quality metric algorithms used [CAND90].

To produce the OPR report, information is obtained as follows: CR and SMR status from the Change Request Tracking System (CRTS), testcase status from the planning tool, software data from the build system, and release date information and release history from internal OPR files (see FIG. D-7).

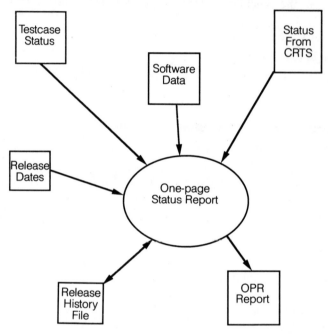

Fig. D-7. OPR process.

Detailed sections

The one-page report is shown in FIG. D-8. The report is divided into six sections, as follows:

1. *Critical Dates.* The Critical Dates section presents all scheduled dates associated with the release and information about the test period. It includes:
   - Report Date—the date the report was run,
   - Scheduled Start Date—the date that testing was started,

```
Report Date: 03-05-91                    ONE PAGE REPORT                      Release:1AM6.2

Scheduled Start Date : 01-22-91      TEST Days Scheduled: 44
Scheduled Freeze Date: 03-11-91      TEST Days in Test : 30
Sign Off Date        : 03-25-91      TEST Days Remaining : 14        CRTS Date : 03-05-91
Ship Date            : 03-27-91      % Days Remaining : 32

****************************************************************************************

                              CONTENT SUMMARY

        ON-LINE                                          BATCH

       SMRs  SMR Items      IR Count                SMRs  SMR Items      IR Count
                       EVAL TEST:   0                               EVAL TEST:  53
Total: 102   224       MUNIT TST:   3       Total:  70   884        MUNIT TST:   0
ENH :   18    61       PROD TEST: 148       ENH :   10   325        PROD TEST:  53
SW  :   79   158                            SW  :   57   537
PROP:    5     5                            PROP:    3    22
DOC :    0     0                            DOC :    0

SMR Qual Index : 30.8     IR Qual Index: 81.8      SMR Qual Index : 56.6    IR Qual Index: 88.7

****************************************************************************************

                               CONTENT STATUS

        ON-LINE                                          BATCH

SMR Item Total: 224     IR Total   :               SMR Item Total : 884     IR Total  : 884
   Closed :  81            Closed :                    Closed : 583             Closed :  53
   Open   : 143            Open                        Open   : 301             Open   :  47
     SW DEV :   7            DEV/STF                      SW DEV :  11             DEV/STF :  6
     SW STF :   1             C:  0 H: 7                  SW STF :   2              C: 0 H: 0
     SW ST  :  69             M:  2 L: 0                  SW ST  : 272              M: 4 L: 0
     SW STI :  60            STI                          SW STI :   1            STI
     DOC DEV:   6            ST                           DOC DEV:   0            ST
     DOC STF:   0            DOC DEV/STF:                 DOC STF:   0            DOC DEV/STF:
     DOC ST :   0            DOC ST                       DOC ST :   0            DOC ST
     DOC STI:   0            DOC STI                      DOC STI:   0            DOC STI
   NT/DEF :  12              CANCEL  :                   NT/DEF :  83              CANCEL  :  5
   CANCEL :  33              DEF/ROL                     CANCEL :  92              DEF/ROL :  0

****************************************************************************************

                          QUALITY METRIC ANALYSIS

Total Est. IRs  : 191    Total Act. IRs   : 201    Est. Kline of Code: 38.440   Act. Kline of Code: 33.119
   On-line      : 122       On-line       : 148    Est. IR Rate /wk  : 18.13    Act. IR Rate /wk  : 33.50
   Batch        :  69       Batch         :  53    Est. Cumm FFD     : 4.120    Act. Cumm FFD     : 0.000
Est. IR as of Today: 130  Act. IR as of Today: 201 Target DD         : 3.000    Act. DD           : 5.43
                          IRs Found Yesterday:   1  Past 5 day DR Rate: 15.00    Production Months :  000

****************************************************************************************

         REGRESSION SUMMARY                            QUALITY INDICES

Total Test Programs: 882    Percentages       QUALITY INDEX  REL: 6.0   REL: 6.1   REL: 6.2
RUN  NO ERROR   : 880         99.8%           Regression      98.53      98.24      96.03
     WITH ERROR : 847         96.3%           SMR             89.70      97.70      51.35
     Batch      :  33          3.8%           IR              79.40      87.00      83.58
     NOT YET RUN:   2          0.2%           RELEASE         85.70      94.50      54.07
```

Fig. D-8. One-page report.

- Scheduled Freeze Date—the date that changes will no longer be accepted,
- Sign Off Date—the date testing is scheduled to end,
- Ship Date—the date the system is scheduled to ship,
- TEST Days Scheduled—the number of working days in the test period,
- TEST Days in Test—the number of working days the system has been in test,
- TEST Days Remaining—the number of working days still remaining in the test period,
- Days Remaining—days remaining divided by days scheduled times 100,
- Release—the system name and level, and
- CRTS Date—the date the information on the report was obtained from CRTS.

2. *Content Summary.* The Content Summary presents information about' SMRs and IRs and is divided into on-line and batch sections and both sections are further subdivided.

   The two content summary sections are composed of three segments, where two of them are related to SMRs (SMRs and SMR items) and the other to IRs. The entries for the on-line section (the batch section is identical but refers to batch software) are:

- SMRs Total—the number of SMRs in the release.
- SMRs ENH—the number of enhancement SMRs in the release.
- SMRs SW—the number of software SMRs in the release.
- SMRs PROP—the number of propagation SMRs in the release.
- SMRs DOC—the number of documentation SMRs in the release.
- SMR Items—the entries are numbers of items, where the items are the separately testable parts of an SMR.
- IR Count—the number of IRs opened.
- EVAL TEST—the number of IRs opened by evaluation test.
- MUNIT TST—the number of IRs opened by multi-unit test.
- PROD TEST—the number of IRs opened by product test.
- SMR Qual Index—the index calculated using the SMR items (testable units) information. From the closed SMR items, subtract the SMR items not tested (NT) and deferred (DEF), and divide by total SMR items (see Quality Metric Algorithms).
- IR Qual Index—the index calculated from the closed IRs, subtracting the deferred (DEF) or rolled IRs (ROL), and dividing by total IRs (see Quality Metric Algorithms).

3. *Content Status.* The Content Status section contains more detailed information about the SMR items and IRs. In this section, the software (SW) and documentation (DOC) SMR items and IRs are classified by open status. The open status values are: development (DEV), system test failed (STF), passed preliminary testing (STI), and system test

(ST). The closed status values are: system test passed (STP), system test resolved (STR), not testable (NT), deferred (DEF), canceled (CAN), and rolled (ROL). The entries for the on-line section (the batch section is identical) are:

- SMR Item Total—the total number of SMR items.
- Closed—the number of items closed.
- Open—the number of items open, with a breakdown just below.
- SW DEV—the number of software items that are in development.
- SW STF—the number of items that were failed by test and not fixed.
- SW ST—the number of items that are to be tested.
- SW STI—the number of items that have been initially tested but require more testing,
- DOC—the same breakdown for documentation items.
- NT/DEF—the number of items that were not testable or were deferred.
- CANCEL—the number of items that were canceled.

The IR subsection of the on-line section contains some additional entries:

- C—the number of IRs with priority critical.
- H—the number of IRs with priority high.
- M—the number of IRs with priority medium.
- L—the number of IRs with priority low.
- DEF/ROL—the number of IRs that have been deferred or rolled to a future release.

4. *Quality Metric Analysis.* The Quality Metric Analysis section presents a clear status of the release currently under test. It places the estimated values for the end of the release side-by-side with the actual ones. The actual values presented are batch and on-line totals unless noted. This section is divided in two groups: estimated and actual metrics.

Estimated measurements are calculated using the least square method that uses historical release data contained in internal files. [OLAG90] The estimates contained in this section are:

- Total Est. IRs—expected number of IRs.
- On-line—expected number of on-line IRs.
- Batch—expected number of batch IRs.
- Est. IR as of Today—the expected number of IRs that should have been opened by the current date.
- Est. lines of code—estimated lines of code in 1000s.
- Est. IR rate/wk—expected number of IRs to be opened per week.
- Est. cumm FFD—expected cumulative field fault density (expected field faults/actual KNCNCSL.)

The actual measurements contained in this section are:

- Total Act. IRs—the number of IRs opened from the beginning of the test period.
- On-line—the number of on-line IRs opened from the beginning of the test period.

- Batch—the number of batch IRs opened from the beginning of the test period.
- Act. IR as of Today—the number of IRs opened from the beginning of the test period (note comparison with estimated IRs as of Today).
- Act. Kline of Code—the actual Klines of code is calculated by adding the new and changed lines of code together and dividing by 1,000. The number of new and changed non-commented source lines (NCNCSL) of code is obtained from the build tool. Actual Klines of Code is an important metric because many of the estimated values have a strong correlation with it.
- Act. IR Rate /Wk—the actual IR rate per week is calculated by adding the total IRs and dividing by days in test divided by five because only working days are considered.
- Act. Cumm FFD—the actual cumulative field fault density is calculated by summing together all maintenance SMRs opened by customers and dividing by the total number of Klines of code. This number will be 0 until the release is shipped.
- Act. DD—the actual defect density is a metric that shows the number of faults found per 1,000 lines of code. It is calculated by adding the total IRs, subtracting cancelled IRs, and dividing by the actual klines of code.
- IRs Found Yesterday—the number of IRs opened during the past 24 hours.
- Past 5 Day IR Rate—obtained by subtracting the actual IRs from five days ago from the current date's total actual IRs and dividing by five.
- Production Months—the number of months that the system has been in production calculated by adding the months for each site. This number is 0 until the release is shipped.

5. *Regression Summary.* The Regression Summary section provides information about the status of automated testcases in regression. Regression is an excellent indicator of quality. The entries are:
   - Total Test Programs—the number of test programs available.
   - RUN—the number of test programs that have been executed from the beginning of the test period.
   - NO ERROR—the number of test programs run that reported no error.
   - WITH ERROR—the number of test programs run that reported one or more errors.
   - NOT YET RUN—the number of test programs not yet executed.
   - Percentages—the percentages are the number reported divided by the total.

6. *Quality Indices.* The Quality Indices section contains a comparison of quality measurements between the current release and two previous releases of a system of the same type. The indices are calculated using

the Content Status and Regression Summary data. The equations are shown in FIG. D-8. The one entry not computed using the equations is:

- QUALITY INDEX—the release used for the comparison.

Quality metric algorithms

The algorithms mentioned in the previous section that are used to produce the one-page report are as follows:

$$Actual\ DIR\ rate\ per\ week = \frac{Total\ OL\ IR\ +\ Total\ BA\ IR}{Days\ in\ test} \times 5$$

$$Actual\ Defect\ Density = \frac{(Total\ OL\ IR\ +\ Total\ BA\ IR)\ -\ (Cancel\ OL\ IR\ +\ Cancel\ BA\ IR)}{K\ lines\ of\ code}$$

$$SMR\ Quality\ Index = \frac{(Closed\ OL\ SMR\ items\ +\ Closed\ BA\ SMR\ items)\ -\ (NT\ \&\ DEF\ OL\ SMR\ items\ +\ NT\ \&\ DEF\ BA\ SMR\ items)}{(Total\ OL\ SMR\ items\ +\ Total\ BA\ SMR\ items)}$$

$$IR\ Quality\ Index = \frac{(Closed\ OL\ IR\ +\ Closed\ BA\ IR)\ -\ (ROL\ \&\ DEF\ OL\ JR\ +\ ROL\ \&\ DEF\ BA\ IR)}{(Total\ OL\ IR\ +\ Total\ BA\ IR)}$$

$$Regression\ Quality\ Index = \frac{Testcase\ run\ without\ error}{Total\ testcase}$$

$$Release\ Quality\ Index = \frac{(Closed\ OL\ \&\ BA\ SMR\ items\ +\ Closed\ OL\ \&\ BA\ IR)\ -\ (DEF\ OL\ \&\ BA\ SMR\ items\ +\ DEF\ OL\ \&\ BA\ IR)}{Total\ OL\ \&\ BA\ SMR\ items\ +\ Total\ OL\ \&\ BA\ IR} \times Regression\ Quality\ Index$$

Where:

IR	=	incident report
OL	=	on-line
BA	=	batch
SMR	=	software modification request
NT	=	not testable
DEF	=	deferred
ROL	=	rolled

Bibliography

ACKE89   Ackerman, A. F., Buchwald, L. S., Lewski, F. H., Software
            Inspections: An Effective Verification Process, IEEE
            Software, May 1989.

BEIZ83   Beizer, B. Software Testing Techniques. New York: Van Nos-
            trand Reinholt, 1983.

BEIZ84   _____. Software System Testing and Quality Assurance. New
            York: Van Nostrand Reinholt, 1984.

BEIZ87   _____. When to Stop Testing. Presented at 1987 National Con-
            ference on Software Testing. Sponsored by Quality Assur-
            ance Institute, November, 1987.

BEIZ90   _____. Software Testing Techniques 2nd ed. New York: Van
            Nostrand Reinholt, 1990.

BOND90   Bond, M. T., House, C. E., Iacovelli, M., Marks, D. M., Rey-
            nolds, L. D., Smith, L.W. Change Request Tracking Sys-
            tem Requirements. Bell Communications Research,
            March 5, 1990; software engineering paper.

BRAD88   Bradley, B. M., Donnelly, K. F., Grey, R. C., Marks, D. M.,
            Patten, M. T., Smith, L. W., TM-STS-012817 Software
            Product Test Methodology Standard. Bell Communica-
            tions Research, October 26, 1988; software engineering
            paper.

BRAN80   Branstad, M. A., Cherniavsky, J. C., Adrion, W. R. Validation,
            Verification, and Testing for the Individual Programmer.
            Special Publication 500-56. Rockville, MD: National
            Bureau of Standards, 1980.

BRUH88   Bruhin, H., Duchnowski, P., Jarosiewicz, O. TM-STS-012516
            ASTRA Interface Specifications, November, 15, 1988;
            software engineering paper.

CAND90   Candelario, A. DREAM Process. Bell Communications Research, November 27, 1990; software engineering paper.

CRAI86   R. D. Craig, Making Independent Testing Work, Proceedings of Third DPMA National Conference on Testing Computer Software, 1986.

DALA90   Dalal, S. R., Mallows, C. L. Some Graphical Aids for Deciding When to Stop Testing Software. IEEE Journal on Selected Areas in Communications, Vol 8, No 2., February 1990.

DEMI87   DeMillo, R. A., McCracken, W. M., Passafiume, J. F. Software Testing and Evaluation. Benjamin/Cummings, Menlo Park, CA, 1987.

DEUT82   Deutsch, M. S. Software Verification and Validation. Englewood Cliffs, NJ: Prentice-Hall, 1982.

DONN87   Donnelly, K. F., Gluck, K. A. pastel and astra: test environments for today and tomorrow. Software Engineering Notes, Vol 13, Issue 1, January 1988.

DOWN87   Downs, T. Reliability problems in software engineering - a review. Computer System Science and Engineering, Vol 2, No 3. Butterworth: July, 1987.

DUNC87   Duncan, M. Assertion Testing. Proceeding of Fourth International Conference on Testing Computer Software. Sponsored by Data Processing Management Association Educational Foundation, June 1987.

GELP88   Gelperin, D., Hetzel, B. The Growth of Software Testing. Communications of ACM, Vol 31, No 6. June, 1988.

GOOD75   Goodenough, J. B., Gerhart, S. L. Toward a theory of test data selection. IEEE Transactions on Software Engineering SE-1: 156-73, 1975.

HETZ84   Hetzel, W. The Complete Guide to Software Testing. Wellesley, MA: QED, 1984.

HIER6   Hiering, V. S., Bennett, D. A. A Developer's Perspective on Software Quality Metrics. IEEE Communications Magazine, Vol 24, no 9:6-11. IEEE Press, September 1986.

HOWD87   Howden, W. E. Functional Program Testing and Analysis. New York: McGraw-Hill, 1987.

JACK79   Jackson, M. A. Principles of Program Design, Academic Press, New York, 1979.

LIND88   Lindgren, A. L., Software Quality Control. Bell Communications Research, December 1, 1988; software engineering lecture.

MANN78   Manna, Z., Waldinger, R., The Logic of Computer Programming. IEEE Transactions on Software Engineering SE-4, 1978.

MARK85   Marks, D. M. TM-ISD-000869 Measuring Test Coverage of
            Software. Bell Laboratories, February 28, 1985; software
            engineering paper.

MARK87   _____. USER requirements for the QUality Evaluation Sys-
            Tem (QUEST). Bell Communications Research, February
            28, 1987; software engineering paper.

MARK87A   _____. TM-STS-009284 Levels of System Test Coverage. Bell
            Communications Research, July 30, 1987; software engi-
            neering paper.

MARK88   _____. TM-STS-011966 Controlling the System Test Process.
            Bell Communications Research, June 22, 1988; software
            engineering paper.

MARK89   _____. TM-STS-014171 Test Procedures for System Test. Bell
            Communications Research, May 12, 1989; software engi-
            neering paper.

MCCA82   McCabe, T.J., Schulmeyer, G.G. System Testing Aided by
            Structured Analysis (A Practical Experience). Proceed-
            ings of IEEE COMPSAC:522-28, November 1982.

MILL70   Mills, H. D. Top Down Programming in Large Systems, in
            Debugging Techniques in Large Systems, Prentice-Hall,
            Englewood Cliffs, NJ, 1970.

MILL81   Miller, E. Howden, W. E. Tutorial: Software Testing & Valida-
            tion Techniques Second Edition,. New York: IEEE Press,
            1981.

MILL86   Miller, E. Using Tools to Improve Test Effectiveness, Pre-
            sented at 1986 National Conference on Software Testing.
            Sponsored by DPMA, September, 1986.

MUSA87   Musa, J. D., Iannino, A., and Okumoto, K. Software Reliabil-
            ity: Measurement, Prediction, Application. New York:
            McGraw-Hill, 1987.

MYER79   Myers, G. J. The Art of Software Testing. New York: Wiley,
            1979.

OLAG90   Olagunja, A. Enhancement of Quality Estimation for
            TIRKS(R) Releases. Bell Communications Research,
            August 9, 1990; software engineering paper.

OSTR88   Ostrand, T. J., Balcer, M. J. The Category-Partition Method
            for Specifying and Generating Functional Tests. Com-
            munications of ACM, Vol 3, No. 6:676-86, June 1988.

PATT82   Patton, G. C. A Standard For Code Counting, Bell Telephone
            Laboratories, October 14, 1982; memorandum for file.

PERR77   Perry, W. Effective Methods of EDP Quality Assurance. Wel-
            lesly, MA: QED, 1977.

PERR83   _____. A Structured Approach to Systems Testing. Wellesly,
            MA: QED, 1983.

POWE82   Powell, P. B., editor. Planning of Software Validation, Verifica-
              tion, and Testing. Special Publication 500-98. Rockville,
              MD: National Bureau of Standards, 1982.

PETS85   Petschenik, N. H. Practical Priorities in System Testing.
              IEEE Software Vol 2, No. 5:18-23. New York: IEEE
              Press, September 1985.

POST88   Poston, R. Preventing most-probable errors in testing. IEEE
              Software: 86-8 IEEE Press, March 1988.

POST88A   _____. Software Engineering SE351I, at Bellcore, May 1988.

RAMA75   Ramamoorthy, C. V., and Ho, S. F. Testing large software with
              automated evaluation systems. IEEE Transactions on
              Software Engineering SE-1:46-58. New York: IEEE
              Press, 1975.

REDW83   Redwine, S. T. An Engineering Approach to Software Test
              Data Design. IEEE Transactions on Software Engineer-
              ing. Vol 9:191-200, IEEE Press, 1983.

WEYU88   Weyuker, E. J. The Evaluation of Program-Based Software
              Test Data Adequacy Criteria. Communications of ACM,
              Vol 31, No 6. June, 1988.

WHIT80   White, L. J., and Cohen, E. I. A Domain Strategy for Com-
              puter Program Testing. IEEE Transaction on Software
              Engineering SE-6:247-57. New York: IEEE Press, 1980.

YOUR79   Yourdon, E., and Constantine, L. Structured Design, Prentice-
              Hall, Englewood Cliffs, NJ, 1979.

YOUR82   Yourdon, E. Managing the System Life Cycle. New York:
              Yourdon Press, 1982.

Index